URBAN AGRICULTURE

NIPA® GENX ELECTRONIC RESOURCES & SOLUTIONS P. LTD.
New Delhi-110 034

About the Authors

Dr. Subas Chandra Sahoo is presently working as Research Officer in the Directorate of Planning, Monitoring and Evaluation under Odisha University of Agriculture and Technology, Bhubaneswar, Odisha. He has Ph.D. degree in Agronomy and working in the cadre of Professor. He has rich experience in teaching, research and extension activities for more than three decades. He has vast experience in maize research, specifically baby corn and sweet corn. He has guided four students of M.Sc. (Ag) in Agronomy. He has more than 35 research papers, 25 abstracts published during various seminars & workshops, two book chapters and one book. He has more than 20 books and 100 popular articles in Odia language for the benefit of farmers and general public.

Dr. (Mrs.) Meenakshi Mohanty is presently working as Nodal Officer, ICAR-Nodal Cell under Odisha University of Agriculture and Technology, Bhubaneswar. She has done her Ph.D. in Agronomy and involved in teaching at under graduate and post graduate level. She has more than 25 years of experience in research and teaching activities. She has research experience on integrated farming system and sugarcane cultivation. She has guided two students of M.Sc. (Ag) in Agronomy. She has more than 25 research papers published in peer reviewed journals, four book chapters, one book and 25 abstracts published in national & international seminar/workshops.

URBAN AGRICULTURE

Subas Sahoo
Directorate of Planning, Monitoring and Evaluation
Odisha University of Agriculture and Technology
Bhubaneswar, Odisha-751003

Meenakshi Mohanty
Officer, ICAR - Nodal Cell
Odisha University of Agriculture and Technology
Bhubaneswar, Odisha-751003

NIPA® GENX ELECTRONIC RESOURCES & SOLUTIONS P. LTD.
New Delhi-110 034

NIPA® GENX ELECTRONIC RESOURCES & SOLUTIONS P. LTD.

101,103, Vikas Surya Plaza, CU Block
L.S.C.Market, Pitam Pura, New Delhi-110 034
Ph : +91 11 27341616, 27341717, 27341718
E-mail: newindiapublishingagency@gmail.com
www: www.nipabooks.com
For customer assistance, please contact
Phone: + 91-11- 27 34 17 17
Fax: + 91-11- 27 34 16 16
E-Mail: feedbacks@nipabooks.com

ISBN: 978-81-19002-37-5

Composed and Designed by NIPA®.

Preface

Increasing interest for healthy lifestyle and inclination towards organic food among the urban people have encouraged them to go for agricultural activities in the form of growing fruits, flowers, vegetables, preparation of value added products, etc. Besides, putting a bed or two of mushroom in the home garden or rearing honey bees in small scale have become popular in urban dwellings. Some of the challenges of urban ecosystem are scanty land resource, high cost of labour, pollution due to vehicular traffic & industrial activities, inadequate sunlight, rising temperature, etc. Utilization of leisure hours of family members, use of household organic waste, production of fresh vegetables & fruits and obtaining a soothing environment are some of the potentialities of urban agriculture.

Due to variation in physical provisions and resource availability, it is necessary to precisely modify the use of inputs and various operations for growing plants and adopting other vocations. In this book, an attempt has been made to compile various recommendations suitable for adoption under urban environment. The recommendations are intended to address the requirement of individual plant or small area of land. General recommendations have been indicated in the book to provide basic idea about growing various plants. However, the dose of organic manure, seed, fertilizer, pesticide and other inputs may vary depending on local climatic situation, season of growing, variety selected, irrigation availability, personal preferences and other factors. This may be customized looking at prevailing situation and resource availability.

It is expected that contents of the book will benefit the students, faculties, urban inhabitants and those interested in urban agricultural activities.

Authors

Contents

1

Introduction

Growth of urban population has become a challenge throughout the world. Presently, many of the cities are featured with high rising buildings, rush of vehicular traffic and networking of industrial establishments. The urban ecological footprints are increasing day-by-day resulting in unfavourable living conditions. The cities require vast area of land to sustain the population for requirement of food, clothing and other requirements. All these items are brought from areas outside the city, may be from far or near. Creating avenue for production of food and other materials in and around the cities will certainly minimize the pressure on land available outside the city areas. The urban sustainability is at threat due to environmental pollution, rising population and intensification of non-green activities. Many cities have natural resources such as water bodies, inherent vegetation, open space, etc.; which can be appropriately utilized for productive purpose. Promotion of urban agriculture in the form of growing crops, planting trees rearing animals and encouraging landscape garden will have long term impact on urban ecosystem. This can prevent environmental degradation, ensure supply of fresh food and improvement in urban life.

Urban agriculture is the method of taking up agriculture and allied activities by utilizing available space and other resources in urban areas. The aim is to use the space for gardening in order to enhance the aesthetic beauty of the immediate vicinity and get economic benefit. It can be used for growing plants for production of fruits, vegetables, greens, flowers and ornamental plants. Besides, it can generate job opportunities, promote greenery in the immediate environment, improve urban landscape and help in waste recycling. People in urban dwellings have developed interest in urban agriculture due to ever increasing environmental pollution and lifestyle stress.

Urban agriculture is also known as urban farming or urban gardening. It deals with growing plants and trees for economic gain, improving aesthetic value and restoration of urban ecosystem. It should envisage distinctive architecture of growing plants under specific situations in the urban areas. In broader sense, it can include rearing animal, mushroom production, bee keeping, etc. Besides, value addition of fruits & vegetables and processing of agricultural produce can also be included in urban agriculture system. Agricultural activities taken up in the urban and peri-urban areas can come under urban agriculture. It can be of recreation type or of small scale commercial enterprises at household level. In many cases, urban agriculture is adopted to meet the daily requirement of the family. In the present day context, consciousness for good quality agricultural produce has motivated the urban dwellers to adopt agricultural practices to produce items for home consumption. Many persons in the urban areas take up agricultural activities to keep themselves physically fit and effective utilization of leisure time.

Urban agriculture is practised in a variety of place including public land, road side area, private garden, park, roof top, terrace, living room, etc. Irrespective of the location, urban agriculture can be an integral part of ecological system of the cities. The design and selection of components depend on the objective and availability of space. Growing vegetables or flowers are seasonal in nature, whereas plantation of fruit trees or avenue trees and development of lawn are considered as long term activities.

There is conspicuous difference between normal agriculture and urban agriculture due to urban regulations, complexity in land, market demand, labour cost and most importantly its effect on environments human health. The prevailing laws in urban areas, sometimes prevent rearing of animals or growing trees at certain places. The increasing market demand for high value crops like celery, broccoli, baby corn, sweet corn, capsicum, strawberry, medicinal herbs, off season vegetables and cut flowers motivates the city dwellers to grow these plants. Sometimes, the open space near the building is meagre and devoid of sunlight due to adjacent high rising buildings. It limits the choice of growing plants in that space. Availability of skilled manpower for agricultural activities is a constraint under urban eco-system. It discourages the entrepreneurs to take up agricultural activities on commercial basis in the cities.

Challenges in Urban Environment

There are several challenges in urban areas due to sewage disposal, municipal waste management, inadequate open space, pressure on water supply, air & water pollution and ill effects of public transport system. Rapid increase in human population and development of economic activities has accelerated urbanization, which often lead to environmental crisis such as air & water pollution, toxicity arising out of industrial & automobile effluents and emissions. It is required to understand the environmental challenges and plan suitable measures to address the issues. Creation of grassland, growing ornamental plants, increasing crop area, development of forest, etc. can promote green coverage in urban ecosystem. The life and life style in urban areas can be improved with adoption of agri-horticultural activities.

Air pollution: Increase in vehicular traffic and industrial activities have intensified air pollution in the cities. Various pollutants such as carbon dioxide, methane, dust, smoke, etc. continuously degrade the air quality of the cities. The air pollution levels in some cities have attained alarming stage, which has strong adverse effect on human health. Extensive plantation and crop related activities are necessary to purify the air and provide a clean environment.

Water pollution: Enhanced urban population, rapid industrialization and intensification of commercial activities have direct bearing on degradation of surface as well as groundwater in the urban areas. The quality of groundwater is adversely affected due to presence of sewerage water, industrial waste and medical effluents. The major causes of water pollution are waste water from industries & commercial establishments, medical waste, effluents from slum area, runoff water, petroleum hydrocarbons, pathogenic microorganisms, toxic substances, sediments from construction sites, pet wastes, sewerage discharge, etc.

Scarcity of water: Water is an economic good in the urban area. The households depend on municipal authorities to meet the water requirement. In many cities, water meters are installed to calculate the water rent depending on volume of consumption. Under such situations, agricultural activities will be costlier by using the priced water. However, plants can be grown by utilizing the waste water from kitchen and bathrooms. Plants requiring less water can be promoted to minimize use of water.

Over exploitation of groundwater: Over exploitation of groundwater has resulted in lowering of water level in cities. The infiltration of water is reduced due to large network of cemented or black top roads, buildings and other infrastructures. Minimum availability of open space, water body and vegetation reduces penetration of runoff water in the soil. Increase in vegetation through growing of crops, gardening, plantation, etc. will accelerate infiltration of water in the soil. This will ultimately improve the groundwater status.

Inadequate sunlight: Urban architecture is dominated by high rising buildings located in close proximity, resulting in reduced availability of sunlight. Closely spaced houses obstruct adequate sunlight to reach the ground level. This has adverse effect on health of plant and human being. Several diseases of plant and animal are attributed to low level of exposure to sunlight. Very often, there is complaint of various disorders because of vitamin D deficiency caused due to absence of adequate sunlight. Low level of solar radiation in slum areas is the cause of pathogenic growth, resulting in spread of harmful diseases. Low level of available sunlight also restricts adoption of suitable agricultural practices.

Rising temperature: In urban areas, there is a common phenomenon of rise in temperature, mainly due to the network of black top roads, densely located concrete structures, heavy rush of vehicular traffic, presence of industrial belts and inadequate greenery. Use of air conditioners in domestic and commercial establishments adds to phenomenal rise in temperature. Increase in temperature is the cause of health problems of human & animal, degradation in living conditions and deterioration of heritage sites. Adoption of agricultural practices in and around urban areas can arrest rise of temperature.

Land features: In the cities, per capita land availability is minimum due to enhanced density of population. There is restriction inland utilization pattern due to regulations imposed by the civic authorities. The demand for construction of buildings, roads and other infrastructure is quite high, which results in minimum availability of land for agricultural activities. Inadequate availability of land compels the urban residents to grow plants on the roof top, terrace and even inside the living rooms. Land availability also influences choice of plant species.

Lack of social coherence: There is variation in social status, economic condition and cultural activities of the families residing in urban areas. It is reflected in choice of growing plants and taking up economic activities in urban colonies. Such diversity in socio-economic characteristics influences selection of crops, taking up horticultural activities and choosing other vocations. For example, a family may prefer growing rose, whereas others may like to grow papaya in the same piece of land.

Costly manpower: Availability of diversified employment opportunity in the cities makes human labour costlier. The labourers prefer to work in industries and commercial establishments for higher income. Hence, it becomes difficult to get the labourers for agricultural work. Besides, the labourers skilled in agricultural operations are not adequately available in urban areas.

Restricting urban laws: The civic authorities of urban areas sometimes discourage use of land for certain agricultural activities. In many cities, it is an offence to rear animals within the residential colonies. There is restriction for growing trees that may come in contact with overhead electric, telephone or cable connections.

Scope of urban agriculture

There are several opportunities in the urban eco-system to grow crops, rear animals and adopt allied vocations. The prevailing market demand for agri-produces motivates the urban inhabitants to go for crop cultivation or secondary agricultural activities. In the recent time, women self-help groups are taking up various activities related to value addition, processing and packaging of agricultural produce.

Potential market: High density of population with more buying capacity generates demand for agri-horticultural produces in the cities. The demand is fulfilled through flow of agri-products from rural areas to urban markets. The potentiality of such urban markets can be harnessed by promoting cultivation of vegetables, flowers, fruits, etc. This will enable the urban people to get good quality vegetables and fruits grown in the backyards or rooftops.

Transportation network: The transportation systems in urban areas are well organised with road, rail and air connectivity. It helps in easy availability of agri-inputs and timely disposal of the produces to distant markets. High value and highly perishable agri-produces can reach the consumer with a short span of time before getting damaged.

Technological support: Generally, most of the public sector research institutes and R&D laboratories of private companies are located in and around the cities. The city dwellers can access technical knowledge from these organisations for growing crops, rearing animals or taking up other activities. For example, several agricultural institutes along with various seed producing companies are stationed in Hyderabad.

Higher education level: The urban population are better educated compared with their rural counterparts. The level of technology adoption is more in case of the educated mass. Hence, adoption of innovative and emerging farm technologies is comparatively higher in the urban areas. The urban entrepreneurs are more inclined towards new generation technologies such as hydroponics, bio-flock pisciculture, terrace gardening, growing high value crops, etc.

Institutional support: There is wide scope for access to institutional support through Government extension machineries, processing & value addition industries, manufacturing units for agri-inputs, farm machinery, etc. An entrepreneur in urban area can have easy access to the inputs and modern technologies available in the city limits.

Availability of organic waste: There is production of huge quantities of organic waste in the form of kitchen by-products, leftover foods from restaurants & hostels, wastes from fruit & vegetable market, dry leaves from avenue plantation & parks, human excreta, etc. These can be effectively utilized for production of organic manures those can be used in agricultural activities.

Migration of labourer from rural areas: There is continuous migration of labourers from rural areas to urban areas in search of livelihood and higher income. This has resulted in higher availability of labour force in urban areas to be engaged in commercial and domestic works. Although the labour cost is high, the domestic workers can be utilized on part time basis for gardening purpose.

Advantages

There are several benefits of growing crops and taking up various agricultural activities in urban area.

Aesthetic value: Urban agriculture has a major objective of fulfilling the aesthetic need of local residents. Many flowering plants like rose, dahlia, chrysanthemum, petunia, aster, marigold, tuberose, gerbera, bougainvillea, etc. are grown to enhance the aesthetic value of the building premises. Several foliage plants like dieffenbachia, aglaonema, croton, areca palm, money plant, etc. are grown to improve the beauty of residential and commercial establishments. Residents having adequate space prefer to maintain lawn for beautifying the house. Now-a-days, lawns are developed on the roof top or in the balcony to add aesthetic value to the house.

Recreation: The people residing in urban areas prefer to spend leisure hours in parks and gardens for recreation. To provide a pleasant experience, the gardens or parks are maintained with green lawns, foliage and colourful flowering plants. Species like *Delonix regia* (*gulmohur*), *deodar*, travellers palm, *kadamba*, etc. are planted in the gardens to enhance the aesthetic value. The home gardens can also be a good place for recreation of the family members.

Avenue plantation: Avenue plantation is an age-old practice in the country to provide shade and fruits to the pedestrians. Tree plantation adjacent to the road network of the city helps to cool down the air temperature and create a favourable environment. Besides, it helps in water infiltration and groundwater recharge.

Reducing family expenditure: In city life, a major portion of daily expenditure is spent for purchase of green groceries and flowers. Urban gardening can minimize the daily expenditure for purchase of fruits, vegetables and flowers.

In addition to this, growing medicinal plants in the backyard/roof top garden can reduce the medicine bill in long run. A family of five needs about 1.75 kg of fresh vegetables per day, a part of which can be obtained by growing vegetables in the house premises. Few plants of papaya, banana, drumstick, lemon, curry leaf, etc. in the backyard can fulfil the household requirement to some extent. Potted plants of marigold, jasmine, hibiscus, oleander, crape jasmine, etc. can meet the day-to-day flower requirement of a family.

Employment generation: Agri-horti activities in urban areas can provide opportunity for family members to get additional income. Floriculture, nursery raising, mushroom cultivation, bee keeping, preservation of fruits & vegetables, preparation of value added products, etc. can be taken up by utilizing leisure hours. The roof top areas can be utilized for mini nursery, mushroom production and bee keeping. This will add to the family income and utilize the surplus time profitably. Moreover, this can generate employment opportunities for the family members and hired labourers in the city areas.

Improving physical and mental health: In the present-day context, the urban dwellers are very much conscious about health and physical fitness. Adoption of urban agricultural practices will ensure physical activities, which can keep people healthy, both physically and mentally. Working for an hour every morning in the garden will result in healthy living. There is adequate mental satisfaction by working in own garden and observing the growth and development of plants. Growing plants in the vicinity of house will increase the oxygen availability, which is a prerequisite for good health. Gardens with green foliage and colourful flowers attract various birds and butterflies. It provides mental satisfaction to the urban dwellers and helps in maintaining biodiversity.

Improving overall environment: Gardening and other agri-horticultural activities around the house improves the overall micro-environment in terms of air purification, groundwater recharge and reducing ambient temperature. Presence of plants augments the aesthetic sense of the surroundings. Many plants have the capacity to absorb toxic substances from air and improve air quality in the immediate vicinity. The micro-water droplets released by the plants help in settling down the dust particles and reduce the ambient temperature. The plant roots play a vital role in enhancing the rate of water infiltration in the soil.

Features and Forms

Urban agriculture can be practised in many forms such as backyard garden, rooftop garden, nutrition garden, terrace garden, etc.

***Backyard garden*:** Backyard garden is the practice of growing vegetables, fruits and flowers in the backyard of the house. It is also known as kitchen garden or nutritional garden. It is mainly done to fulfill the daily requirements of the family and sharing the surplus produce among friends and neighbours. The surplus produce can also be preserved as value added products for future use. The household wastes including kitchen waste, vegetable peels, leftover food, used flowers, etc. are productively utilized as manure in the backyard garden.

***Rooftop gardens*:** Most of the urban houses are short of open space for gardening activities. Hence, the open areas available on the rooftop can be effectively used for gardening purpose. Plant species including vegetables, fruits, flowers, foliages, medicinal plants and other herbs can be taken up in rooftop gardens. Some people use rooftops for developing lawns for family recreation. The plants are mostly grown in pots or raised beds are prepared on polythene sheets. Rooftop gardens minimize ambient temperature and enhance the air quality. Besides, gardens on the rooftop beautify the buildings. However, care should be taken to prevent seepage of roofs or any other damage to the building by use of heavy materials during gardening activities.

***Street landscaping*:** In cities and towns, the sideways of the street are covered with avenue plantations, flowering plants, foliage plants, etc. to make the streets more beautiful and to improve the urban air quality. It also reduces runoff during rainy season, thereby protecting the top soil from erosion. In some places, road side nurseries are developed to provide good quality planting materials to the urban dwellers. *Deodar*, *Gulmohur*, *Kadamba,* travellers' palm, date palm, etc. are planted as avenue plants. Bougainvillea, Yellow oleander *(Kaniara), Nerium (Karabira)*, Thuja, Juniper, etc. are some of the ornamental species grown along the roadside.

***Forest plantation*:** This is related to plantation of forest species in unban ecosystem. This includes tree species such as bamboo, teak, *kadamba*, *Mimusops elengi*, *Cassia fistula* (golden shower), *Neem* tree, *Peepal* tree, *Cassia siamea* (rain tree), etc. Fruit trees including mango, *jamun*, almond, ber, etc. are planted in the open space. This type of plantation provides conducive atmosphere for urban life and promotes biodiversity conservation in the cities. Forest gardening can be a part of afforestation activities, which acts as a step forward for mitigating ill effects of temperature rise in urban areas.

***Green house*:** Green house is a structure made up of glass/ polyethylene/ plastic for providing controlled condition for growing plants. Greenhouse provides facility to grow crops round the year under regulated environment.

Greenhouse protects the plants from extreme temperature during summer and winter seasons. It is established for commercial purpose and needs a sizable area of land and initial investment. High value crops like capsicum, broccoli, celery; offseason vegetables like cabbage, cauliflower, tomato, coriander, radish, etc. (grown during kharif season) and flowers like gerbera, anthurium, petunia, etc. can be taken up as profitable crops in the greenhouse. Though the initial cost of establishment is high, greenhouse can be profitable in long run due to production of high value crops.

***Green walls*:** Green wall is a vertically constructed structure having vegetation cover with an objective to enhance the aesthetic value of the buildings and roads with minimum ground area coverage. Green walls are also known as living walls or vertical gardens and are widely adopted both in outdoors and indoors of hotels, restaurants and places of tourist attraction. Various growing media are placed vertically for raising foliages and flowers.

***Backyard permaculture*:** Permaculture is the agricultural practice, which aims at creation of self-sufficient and sustainable agricultural ecosystems. It depends on utilization of natural resources and develops a synergetic crop production system based on crop diversity and resilience. Permaculture integrates land, water, forest, nature, human being and the environment through mutually beneficial system. Developing a sustainable ecosystem in the backyard is important in urban areas by combining plants and animals suitable for the local climate. Backyard permaculture can promote a healthy environment that provides recreation along with production of healthy food.

***Container garden*:** Container gardening is otherwise known as pot gardening. It is the method of growing plants in containers or pots instead of planting on the ground. The container used in this method is small and usually portable. It can accommodate flowers or other plants. Container gardens are a perfect way to utilize small outdoor space or rooftop for growing plants. Urban farmers can produce fresh vegetables and herbs for the family through container gardens. This type of gardening is very much helpful to the persons facing acute shortage of land area. The extension of container garden with large scale plantation is known as patio garden.

***Polyculture for small gardens*:** This is the combination of growing plants in limited area for obtaining food and other produce. In simple words, polyculture is the combination of plants within a garden to provide several plants at the same time and same space. Polyculture can be adopted by the persons having very little area to grow crop. For example, lettuce can be grown at the base of corn plants. Similarly, climbing plants can be grown on trellis along with bush-type plants. This is an intensive approach to obtain high yields from limited area. In this case, soil should be fertile and additional quantities of nutrient are to be incorporated in the soil to sustain more crops.

Future Strategy

Cities can develop strategies to intensify urban gardening. Some of the strategic points are discussed here to promote urban agriculture activities.

***Addressing existing challenges*:** Urban gardening should be in a position to mitigate various issues relating to urban life. It should be able to provide fresh and healthy fruits, vegetable, milk, etc. for consumption of the growers. The urban environment can be improved through plantation of various trees, growing crops and avenue plantation. The waste material can be profitably disposed by the activity of urban farmers. All these factors are to be taken into consideration while preparing strategy for urban farming.

***Inventory of land and rooftops*:** Although the cities are expanding both horizontally and vertically, there is dearth of space around the houses, public places and commercial establishments. Still, some spaces are available under public and private domain along with rooftops to accommodate agricultural activities. Most of the cities are having public land, private area, rooftops, greenbelts, parks, gardens, etc. Hence, it is necessary to prepare inventory of such available space, those can be utilized for growing agri-horti crops, lawns, avenue plantation and forest plantation. The inventories can be utilized for food production, creation of recreational avenue, improvement in aesthetic value, environment protection and biodiversity preservation.

***Collaboration and cooperation*:** Agricultural activities in urban area involve several stakeholders including community members, input suppliers, market players, transporters, associated institutes, etc. Besides, the urban authorities involved in water supply, town planning and waste disposal have to be taken into confidence prior to taking up urban agricultural activities. Strategizing agri-horti activities in the cities needs to involve these stakeholders. The forward and backward linkages can be strengthened by collaborating with the input supplier and the buyers of agri-produce. Technical support from nearby research institutes can be availed for use of latest technologies to make urban agriculture economically viable. The services of Non-Government Organizations can also be utilized for creation of awareness and motivating the citizens.

Cooperatives and associations: Production of food and other materials can be achieved more effectively through collective effort of all the stakeholders. Producers' cooperatives can be formed in the slum areas to involve the dwellers for food production and employment generation. Cooperative efforts of production and processing of agricultural produce will have a better bargaining power for procurement of input and disposal of produce. The associations and organizations functioning in different parts of the city can be involved in promotion of ornamental horticulture and beautification of residential as well as institutional area. This can also help to fulfil some of the daily requirements of the local residents in terms of fruits, flowers, vegetables, etc. Adoption of cooperative approach in urban agricultural sector helps in protection of the crops and plants. Besides, this can also protect the interest of the producers to fetch fair price and to avoid exploitation by the middlemen.

***Within the purview of law*:** Many cities have strict laws which prohibit indiscriminate use of land within urban areas for plantation of crops, rearing animals and related activities. In some cities, poultry farms are not allowed within the residential area. Similarly, it is not permissible to keep pet animals in some of the high rising buildings. A forest tree should not be planted under high tension electric wires. In urban areas, the land is categorized for definite purpose such as residential area, commercial establishment, institutional area, industrial belt, green zone, heritage corridor, etc., which has restriction on land use pattern. While planning for agricultural activities, the existing law of the area has to be obeyed.

***Market linkage*:** The urban gardeners usually produce for their own consumption or grow plants as a recreational activity. Sometimes, there are surplus produces which need to be marketed. Many entrepreneurs also take agriculture as a profitable vocation for earning their livelihood. Hence, it is necessary to develop linkage with the local market or markets at distant place. There is wide scope for marketing the garden fresh produce through farmers' markets, shopping malls and roadside venders. Procurement of inputs such as seed, fertilizer, organic manure, plant protection chemicals, garden tools, implements, etc. can be done cost effectively by assessing proper market.

***Recycling the resources*:** There is vast production of organic materials in terms of kitchen waste, garden waste, etc. It has become a liability on the part of urban authorities for safe disposal of these waste materials. Appropriate strategy can be developed to utilize these materials for production of food, development of lawn and promoting gardens. This will be beneficial both for the urban gardener and city authorities. The urban environment can be improved by converting the waste material to organic manure that can be used in agri-horti activities.

2

Types of Urban Garden

Urban agriculture involves various aspects of growing plants and taking up other activities. Those include terrace gardening, vertical gardening, green walls, hydroponics, etc. Although several other types of garden present in urban areas, the gardens suitable for residental premises are discussed in chapter.

Terrace Garden

Terrace garden is generally established on a terrace or rooftop of a house having limited land space for gardening. This type of gardening is gaining momentum in the cities. These gardens can be created on different buildings like residential flats, individual houses, public buildings, commercial establishments and factories. The growth and development of the plants give immense pleasure to the garden owner. Plucking a fresh cucumber on the terrace provides a different pleasure, which is beyond the physical value of the fruit. The experience of consuming a fresh watermelon from own garden has unique satisfaction. As there is minimum or no space for growing plants in urban areas, the terrace can be transformed to an attractive garden. Irrespective of size of the terrace, it can be converted to a beautiful garden with proper planning. There are multiple ways for revamping the terrace depending on available time and resources.

Some people grow herbs and vegetables on their terraces, while others grow ornamental plants for aesthetic purpose. Some garden owners prefer to grow all these depending on the availability of space. Plants can be grown on trellis around the terrace for privacy or shade. This can also help to create a pleasant recreation area.

Benefits of terrace garden

- The terrace garden can provide nutritious fresh vegetables and greens. Terrace garden can be a source of good quality organic vegetables and fruits, which are beneficial for health.
- The flowers grown in the terrace add colour to the buildings. This can also fulfil the daily flower requirement of the family.

- The micro-environment gets cooler and the surrounding air is filled with fragrance and enriched with oxygen. This will reduce the energy consumption of the family due to reduction in use of air conditioner.
- Terrace garden will be an ideal place for relaxation for the family members. The children can be acquainted with growth and development of various plants.

Steps associated with terrace gardening

The following steps are to be taken prior to developing a terrace garden.

Layout: Preparation of layout/sketch helps to decide the type of plants, space utilization and investment requirements. A rough sketch may be prepared as per location of the terrace, availability of water source, local weather condition and requirement of the family. The planning should take into account of the available resources *viz.* family labour, investment capacity and knowledge on gardening. Moreover, the method of growing plants such as planting in pots/ containers, on raised beds or on the vertical structures on the terrace should be taken into consideration. If desired, as per availability of space, sitting arrangement can be made in the garden.

Terrace capacity: Before starting terrace garden, assessment may be done whether the roof surface can bear the weight of the soil and plants. Usually, wet soil is much heavier to add load on the terrace. Hence, light soil may be used for growing plants on the terrace. Terrace garden should be avoided on the roof structure showing crack or with structural damage. A terrace garden should be made on the rooftop of a strongly constructed building or a house constructed with load bearing floors.

Equipment or tools: Various tools are needed for gardening activities on the terrace. This will reduce the drudgery and make gardening activities less cumbersome. Some of the commonly used tools and accessories are trowel, shovel, garden rake, pruning scissor, fork, *khurpi*, secateurs, hand sprayer, rose can, hose pipe, etc.

Water proofing: Prior to initiation of terrace gardening, it is essential to protect the terrace from water leakage. When plants are sown in the raised beds/ pots/ containers, there is chance of water seepage through plant root system as well as from the moistened soil mass. This may result in damage of the roof and even entire building. Hence, it is necessary to take adequate steps to protect the building structure, tiles, colour and electrical wiring. Waterproofing can be done through polymer coating or spreading polythene sheets on the floor of the terrace. Minimum water should be applied to the plants to avoid overflowing.

The drainage system should be adequate to drain out excess water from the garden. To prevent clogging of the drainage system, it should be placed separately from the plants.

Soil preparation: It is necessary to prepare a healthy soil mixture with addition of sufficient quantity of organic manures to enrich the soil. Properly nourished and conditioned soil is the key to growing plants. The ideal soil for terrace garden should neither be too compact nor too light. The soil should have a balanced mixture of soil, sand and organic matter. Coco peat can also be used as a substitute to the soil, particularly for the potted plants. Coco peat can retain moisture for longer period and also supply plant nutrients. The soil mixture can be used after one or two weeks of preparation for growing plants. The organic matter may be well decomposed household waste, farm yard manure, urban compost, leaf mould, vermi-compost, etc. Several bio-fertilizers like *Rhizobium* culture, *Azotobacter*, *Azospirillum*, phospho-culture can also be used to supplement plant nutrients. If necessary, minimal use of chemical fertilizers can be done to meet the nutritional requirement of the plants.

Wind barriers: There is risk of damage to the garden by occasional strong wind, especially in the tall buildings/ high rising apartments. Wind barriers in terms of trellises, hedges, fences or shelterbelts can reduce the pressure of the wind and prevent the plants from damage. The direction of the terrace should be considered for preparation of wind barriers. Care should be taken for free circulation of air in the garden for proper plant growth.

Sunshine/shade: There is variation in the intensity and duration of sunlight received on the terraces depending on location of the building and its adjacent structures. The position of the terrace also determines the hours of sunshine and shade period. Selection of plants for terrace garden should be done according to the period of sunlight received on the terrace. Shade loving plants like *Dieffenbachia, Aglonema, Dracaena, Syngonium, Coleus, Anthurium,* money plant, *etc.* can be taken up on the terraces receiving less hours of sunshine. The terraces exposed to more sunshine hours can be used for growing plants suitable under bright light intensity such as areca palm, royal palm, capsicum, tomato, brinjal, broccoli, french bean, cucumber, bitter gourd, cauliflower, cabbage, greens, coriander, chilli, marigold, rose, petunia, dahlia, chrysanthemum, etc. Lawns can be developed on terraces receiving bright sunshine for prolonged period. Salad greens like lettuce, spinach, mint, kale, etc. can grow well in partial shade.

Heat proofing: Terrace garden needs special care during summer months as it is developed on a cemented structure. The plants suffer from wilting due to excess heat stress during summer. The plants should be watered daily to keep the micro-climate cool. Adequate quantity of organic matter in the soil can make the soil environment favourable for root growth. If possible, shade-net can be used around the terrace garden to protect the plants from scorching effect of sun and ill effects of hot wind.

Watering: The soil in terrace garden is shallow and dries up quickly. Therefore, watering is to be done more frequently to keep the plants hydrated and healthy. While watering, minimum quantity of water is to be used in terrace garden. Watering the plants with the help of rose can or drip system can help in economising water use. Watering should preferably be done during morning hours to keep the plants hydrated and support the photosynthetic process. Care should be taken to drain out excess water in the terrace garden.

Pest management: The plants grown in the terrace garden are likely to be affected by various diseases and insects. The source of disease pathogen may be from seed, soil mixture, air and water. Care should be taken to protect the plants from these pests. As far as possible, bio-pesticides and botanicals are to be used for control of insect and disease. Addition of neem or *karanj* oilcake with the soil mixture will address the insect and pathogen originating from the soil. Spraying of neem based pesticide and use of bio-control agents can manage several disease and pest. Sometimes, birds damage the fruits and plants. The fruits can be protected from birds by using bird scarring ribbon or plastic net.

Plants suitable for terrace gardening: Several plants can be grown in the terrace garden. Selection of plants depends on availability of space, investment capacity, availability of water, personal choice, family requirement, availability of sunlight, position of the terrace, load bearing capacity of the structure, etc.

- Terrace having ample space can accommodate lawns, spreading shrubs, fruit trees and variety of plants.
- Plants requiring intensive care and high value crops need more of investment. On the other hand, palms, agave, dieffenbachia, money plant, aloe vera, etc. require minimum investment for plantation and subsequent maintenance.
- Adequate availability of water can ensure plants requiring more water e.g. lily, greens and other foliage plants. Scarcity in water may compel the gardener to go for low water requiring plants like bougainvillea, lantana, etc.
- The choice of plant depends on family need. Some families require more of vegetables, whereas others need more flowers. Vegetables like

brinjal, chili, cucumber, bottle gourd, bitter gourd, ridge gourd, ladies finger, etc. can be grown during rainy and summer season. Cauliflower, cabbage, beans, capsicum, tomato, broccoli, coriander, etc. can be taken up during winter months. To meet the daily requirement of flowers; marigold, jasmine, hibiscus, etc. can be planted. To fulfil the aesthetic requirement; perennial foliage plants, rose, anthurium, bougainvillea, dahlia, chrysanthemum, petunia, dianthus, flux, etc. can be planted.

- In some cases, people prefer to grow medicinal plants like *ocium*, *aloe vera*, *kalmegh*, *aswagandha*, *bramhi*, etc. for family use.
- The terrace receiving adequate sunlight for longer hours can accommodate sun loving plants whereas, shade loving plants can be preferred on terrace receiving less sunshine.
- There is variation in sunlight and wind speed depending on the direction and position of the terrace. Selection of plants may be done accordingly.
- Plants of small stature may be selected for the terrace with low load bearing capacity. Plants with heavy foliage should be avoided in this case. The pots/containers are to be filled with less quantity of pot mixture to lessen the burden.

Flower beds on the terrace

- Small flower beds can be made directly on the rooftop or terrace. This can be done by putting the soil on specifically prepared place.
- Water proof stones or brick works can be used to create inner wall of the flower bed.
- Annual flowering plants such as pansy, antirhinum, alyssum, dianthus, phlox, etc. are suitable for terrace garden.

Minimizing operational cost

Development of a terrace garden requires heavy investment at the initial stage. It can be minimized by following simple steps.

- Reuse of old furniture by converting into containers for growing plants. Old cartons, cardboard boxes, paint boxes, food containers, etc. can be used as container. This can reduce the cost for purchasing separate containers for growing the plants. This will not only cut down the expenditure, but also minimize waste generation in the home premises.
- The organic remains of the house such as fruits peels, vegetable waste, leftover food, used tea leaves, etc. can be used to provide nutrient to the

plants. This will ultimately minimize the cost for purchase of organic manure and fertilizer.

- Growing disease resistant variety, seed treatment and use of neem or karanja oilcake will reduce incidence of disease and pest, which will lower the expenditure on purchase of plant protection chemicals.
- Involvement of family members in all the operations will help to reduce the expenditure on labour for developing and maintenance of terrace garden.
- Planting short duration crops will require less input in terms of water, manure, pesticide and labour. It will ultimately minimize the cost of operation.

Hydroponics

Hydroponics is the promising area of agricultural activity dealing with growing plants in a nutrient solution. It is the process of growing plants in water solution without using soil. In case of nutrient film technique, the roots are suspended in a liquid medium or supported by an inert medium. But, if the roots remain in floating condition, then it is termed as deep floating technique. This technique has been in use since many years as evidenced from various countries. With continuous waning of arable land, emphasis is given on hydroponics to meet the ever increasing food requirement of burgeoning population. Due to extremely scarcity of land in urban areas, hydroponics can be a reliable option to grow crops, ornamental plants and other species. Hydroponics can be taken up in balcony, terrace, rooftop and space available adjacent to the buildings.

Hydroponic system operates by manipulating temperature, plant nutrient, water and environmental condition. The basic principle of hydroponics is to provide the exact need of the plant at proper time. Tailor-made nutrient solutions are given to the plant for growth and development. This system also controls the light intensity and duration precisely as per suitability of the plants. The pH level of the growth medium is monitored as per requirement of the plants grown.

When the plants are grown in controlled environment, many risk factors such as moisture deficit, imbalanced plant nutrition, weed growth, occurrence of disease, incidence of pests, etc. are minimized. Absence of disease, pest, weed, etc. eliminates use of plant protection agro-chemicals, which results in production of good quality fruits and vegetables. Hydroponic systems get rid of the uncertainty associated with growing plants in outdoors. The lifecycle of the plant is fastened, when grown without soil due to absence of mechanical resistance of soil.

In hydroponic system, the plant growth is mainly influenced by the nutrient solution. The success of hydroponic system depends on use of proper quantity of nutrient in right dose and right concentration. The nutrient solution influences the quantity and quality of production. This technique has several benefits over other methods of cultivation with respect to utilization of space, enhanced plant growth, increased water use efficiency and less infestation of disease & pest.

Features of hydroponic system

- The roots are hanged to the nutrient solution, either in the form of liquid or mist.
- The roots grow in the inert medium of sand, gravel, rockwool, perlite, vermiculite, peat moss, foam, coconut coir, etc.
- The nutrient solution is provided from a pool to the plants. It is collected and reused after passing through the root zone.
- It is important for sterilization of solution so that pathogens from one plant cannot spread to other plants.

Advantages of hydroponics

- Plants can be grown without requirement of soil.
- It is helpful in the city areas due to minimum availability of space.
- There is no risk of soil borne disease or pest.
- There is control over the rhizosphere for efficient root growth.
- Nutrient concentration and pH can be adjusted as the roots are grown in the nutrient solution.
- High density planting is possible to minimize the requirement of land area.
- Higher yield per unit space due to intensive management practices.
- Better control over water, nutrients, pH and the environment resulting in higher yield of plants.
- Efficient use of water and nutrients due to less evaporation and minimum leaching loss.
- Better growth and development of plants due to absence of weeds.

Disadvantages of hydroponics

- The initial investment is high for setting up of hydroponic system.
- It requires adequate skill and technical competence to establish and operate the system.
- As the system is cost intensive and needs precise technical competency, high value crops can only be sustainable in the system.
- There is dearth of technology for growing various crops hydroponically.
- It is difficult to manage diseases and insect pests in the crop due to continuous submergence of plant roots in nutrient solution.

Growing media

In hydroponic system, various types of media are used to maintain desired oxygen level and support root growth. The media should be inert and strong enough to support the weight of the plant and provide anchorage to the root.

***Coco peat*:** Coco peat is prepared from coconut husks and acts as a suitable growth medium for hydroponic system. This is a sustainable organic medium and cost effective in nature.

***Expanded clay pellets*:** The clay pellets are made up of expanded clay to form round balls of porous material. These pellets are pH neutral and do not release any nutrients. Besides, the spherical shape and porosity help to ensure good water balance for favourable root growth.

***Rice hull*:** The husks of paddy are used as a medium in hydroponics. This is the by-product of rice that can be effectively used for growing plants in hydroponic system.

Sand: Sand can also be used as a medium in hydroponics. This is the cheapest medium and easily available. But, it adds weight to the system and may not be encouraging to take in the balcony, terrace or roof top.

***Gravel*:** Small gravels are also used to support plant growth in hydroponics. Colourful gravels add beauty to the system and make hydroponics attractive. It is inexpensive and recyclable.

***Brick debris*:** The broken bricks generated during construction work can also be used as a medium for plant growth. Poor quality bricks, which dissolve in water are not desirable for the medium. The bricks may affect the pH of water and require extra cleaning to remove brick dust.

***Wood chips*:** This is the product of saw mills, which can successfully used for growing plants in hydroponics. It is organic in nature and holds the structure for a long period.

***Pumice*:** It is a light weight mineral that is crushed and used as growing medium in some hydroponics systems. It has the capacity to retain high oxygen level.

***Rockwool*:** Rockwool cubes are used for growing plants under hydroponic system. The fibres create a structure suitable to retain water and oxygen.

***Starter plug*:** It is made up of organic compost and doesn't disintegrate like soil due to presence of biodegradable binding material. It is the most convenient growth medium for growing large number of plants. This is a sustainable growth medium for hydroponics.

***Air stones*:** These are used in hydroponics for aeration of water. This is necessary to prevent drowning of plants under submerged condition. Air stones dole out tiny bubbles of dissolved oxygen throughout the nutrient solution. These bubbles help in proper distribution of the dissolved nutrients in the solution.

***Net pots*:** Net pots are the mesh planters used to support the plants in hydroponics. This is the material enabling root growth from the sides and bottom of the pot for better availabilities of oxygen and nutrients. Net pots also facilitate better drainage as compared to other pots.

Types of hydroponic system

There are various methods of hydroponic system. These methods are modification or combination of six basic hydroponic systems.

***Deep water culture hydroponics*:** This is also known as deep film technique. Deep water hydroponic system is the easiest and most popular method of hydroponics. In this system, the plants are suspended in aerated water. The plant roots are submerged in the nutrient solution and get constant access to nutrient, water and oxygen. Deep water culture is known to be the purest form of hydroponics. It is very easy to prepare deep water culture system in terrace or rooftop without using expensive equipment. A clean bucket containing solution with a floating surface like styrofoam can be used to grow plants. The stem or other part of the plant is not placed under the water. This system requires minimum maintenance. It needs to replenish the nutrient solution at the time of need. Oxygen is to be provided intermittently by pumping. This system is not ideal for flowering plants. This method is useful for growing plants with short root system and is comparatively cheaper than other methods. Wick system: In this system, the plants are grown on a tray containing media and kept on the top of a container with nutrient solution. Wicks are dipped in the nutrient solution, which carry the nutrients to the plants. Water and nutrients flow up by the wick and saturate around the plant root systems. The wicks can be made up of rope or string. Wick system is the simplest form of hydroponics, which works on the principle of capillary action. This system is suitable for the fast growing plants requiring less water. As the system is always humid and damp, there is risk of fungal attack and incidence of rotting disease.

***Nutrient film technique system*:** In this system, the nutrient solution is moved through the roots of plants placed in a channel made up of wood or plastic tubes. The nutrient rich solution is passed through by gravitational force or by pumping to reach the plant roots. This technique is effective for the plants having long root system. The roots of the plants are not immersed in water, rather the stream of nutrient solution flows over the tips of the roots. The plants are allowed to grow on tilted channels. The flowing nutrient solution runs down and is collected in a reservoir kept underneath. This water is again passed through air stone to absorb oxygen. Thereafter, the nutrient-rich water is pumped in to the channel by a submersible pump. Thus, the nutrient film technique is a re-circulatory hydroponic system. It is wise to replenish the nutrient solution once in a week to get better results. This system is popular as it can support several plants in the channel. This system requires less quantity of water and nutrient.

***Ebb and flow system*:** This system works on the principle of flooding the bed with nutrient solution from a reservoir and draining out intermittently. The submersible pump fitted with a timer lifts the nutrient solution to the bed with plants. Subsequently when the timer stops, the excess solution is drained back to the reservoir by gravitational force. This is a modified system of alternate flooding and drying. The plants take up the nutrients during flooding and get aeration after drain out of the solution. Ebb and flow systems can accommodate various types of plants such as tomato, pea, beans, cucumber, carrot, capsicum, etc. Trellises can also be attached with this system for growing climbers and creepers. This can be a good system to grow vegetables and flower crops on the rooftop or even in the space available near the buildings.

***Drip system*:** Drip system is the most popular method of hydroponics, mainly adopted by commercial growers. In this case, the nutrient-rich reservoir supplies solution to the rhizosphere of individual plants through a network of tubes to keep the plants moist and nourished. The drip system of hydroponics is designed to deliver nutrient solution precisely to keep the growing media damp. This system can support large-scale hydroponics operations by accommodating large number of plants. Drip system can be taken up in urban areas to produce crops by reducing water requirement.

***Aeroponics*:** In aeroponics system, plants are suspended in the air and roots are exposed to a nutrient containing mist. This is a system with enclosed frameworks such as cubes or towers to hold a multitude of plants. Water and nutrients are pumped from the reservoir to a nozzle, which converts it to fine mist. The mist is usually released from the roof of the chamber and moves down. Constant exposure of the roots to air allows those to take oxygen and grow at an accelerated rate. Aeroponics does not need substrate media to survive. Aeroponics system uses less water as compared with other form of hydroponics. In this system, higher yield can be obtained from limited space. Vine plants, tomato, lettuce, brinjal, etc. can be successfully grown under aeroponic environment. Aeroponics is the eco-friendly hydroponic system. This can be well adopted on the rooftop and terrace garden. There is less risk of damage to the floor due to less use of water. It keeps the micro-environment cooler due to presence of mist. However, the initial investment is higher as compared to other methods of hydroponics.

***Aquaponics*:** The term aquaponics is derived from two words viz. aquaculture and hydroponics. Aquaculture refers to fish farming and hydroponics means growing crops in water. Aquaponics is the technique of growing plants by maintaining symbiotic relationships between plants & fish. The symbiosis is attained as nutrient-rich waste from fish tanks serves as food for hydroponic production system. On the other hand, the hydroponic beds help to remove gases and chemicals. Besides, gravel beds provide habitats for nitrifying bacteria, which improve the nutrient cycling and filter water.

Nutrient solution: Nutrients are the most important factor for plant growth in hydroponics. Supply of nutrient elements depends upon the type of plant, media used, growth stage of plant and the environmental conditions. Nutrient solution is prepared by mixing fertilizer with water in a definite composition in order to support plant growth. The composition of nutrient solution plays a vital role for plant growth. Testing of water should be done before using in hydroponics. The nutrient solution for hydroponics can be prepared by mixing fertilizer with water or by use of premixed solutions. The selection of fertilizer should be such that there would be minimum effect on pH of the solution. The formulations must be based upon the requirement of the targeted crop. The nutrient solution should have desired levels of pH and EC. The nutrient solution passed through the roots is recirculated for further use by plants. In this process, about 50 per cent of the fertilizer and 70 per cent of water are utilized by the plants

Vertical Garden

Vertical farming is the practice of growing plants on vertically inclined surfaces under controlled condition. In this system, plants are grown in vertically stacked layers developed on structures like skyscraper, shipping container or repurposed warehouse. It includes growing the plants in soilless media. In vertical farming, crops are grown indoors under artificial condition of light and temperature. It gives higher production from lesser area. Plants grown in vertical farming requires considerably less water and agro-chemicals as compared to traditional plant growth techniques. Being placed indoor, the plants are not exposed to seasonal variations, hence give high productivity year-round. Some common structures to accommodate vertical farming systems include buildings, tunnels, walls, etc. High value crops should be grown in vertical farms to achieve more monetary benefit as the initial cost of establishment is considerably high.

The artificial control of temperature, light and humidity makes the system more productive. The primary goal of vertical farming is maximizing crops output in a limited space.

Vertical agriculture is an innovative way of maintaining the agricultural practices. In India, vertical farming is mostly polyhouse-based farming. Poly-house farming is a protected way that gives higher productivity from vegetables and fruits

Vertical farming technologies are relatively new to crop production system. It is difficult to produce crops at large scale under vertical farming system. Still it is economically feasible to meet the growing food demand. The technologies developed for vertical farms can be adopted by other indoor farming sector such as greenhouses. It can effectively utilize natural sunlight.

Establishment of vertical farm: Establishment of vertical farm involves understanding basic working principle, physical layout, provision of artificial lighting, selection of appropriate growing medium and taking care of sustainability features. The primary goal of vertical farming is to grow more plants per unit area. In this case, crops are grown in stacked layers. A perfect combination of natural and artificial light is decided to fulfil the light requirement of the plants. Technologies such as rotating beds are used to improve light efficiency.

In vertical farming; aeroponic, aquaponic or hydroponic growing media are used instead of soil. Peat moss or coconut husks and similar non-soil media are common in vertical farming. The vertical farming method should be sustainable economically and physically in terms of lesser energy consumption with higher production in long run.

Food crops can be cultivated easily in urban areas by planting in vertically stacked layers in order to save space, energy and water. The plants are vertically stacked in a tower-like structure. Hence, the area required to grow plants is minimised. A combination of natural lights and artificial lights maintain a perfect environment suitable for growth of the plants.

A vertical farm enable farming within the city area. The produce is quickly delivered and always fresh, when compared with the produce obtained from the distant place. The spoilage during transportation is also reduced.

However, vertical farming has its own demerits. Initial capital cost for establishing vertical farming system is the major constraint. In addition, there is cost involved for erecting the structures along with its automation.

Advantages of vertical farming

Vertical farming has a lot of promises and potentiality for the future. Some of the major benefits of vertical farming are:

- ***Year-round crop production***: Year-round crop production is possible in a controlled indoor environment. Vertical farm products are available near the urban areas and are easy to market.
- ***More production from less area***: Vertical farming allows to produce more crops from limited growing area.
- ***Less use of water***: Vertical farming allows to produce the crops with use of less water as compared with normal cultivation.
- ***Minimum effect of adverse weather***: Crops in a field can be adversely affected by natural calamities such as rain, cyclone, flooding or drought.
- ***Organic production system***: As the crops are produced in controlled indoor environment without use of chemical pesticides, vertical farming allows to grow crops organically.
- ***Environment friendly***: Indoor vertical farming can lessen the occupational hazards associated with traditional farming. Farmers are not exposed to fertilizers, pesticide and toxic chemicals. This method of farming is environment friendly due to low emission of greenhouse gases.
- ***Biodiversity çonservation***: Vertical farming is less harmful to the native plants and animals, thereby promoting biodiversity conservation.
- ***Low water requirement***: Vertical farming requires considerably less water for growing plants.

- ***Lowering temperature***: Indoor vertical farming lowers the ambient temperature of the house.
- ***Plants in layers***: Various types of plants can be grown simultaneously at different layers
- ***Easy in pest management***: Pest and disease management is easier in vertical farms as the farming is done under controlled condition.

Limitations of vertical farming

Vertical farming has some disadvantages, as detailed below.

- ***High initial investment***: Vertical farming technologies require high initial investment compared to traditional farming methods.
- ***Difficulty in pollination***: Vertical farming takes place in a controlled environment without presence of insects. It adersely affects the pollination process needs.
- ***Labour intensive***: Vertical farming requires extra care for management of the plants, which is labour intensive. The cost of labour is substantially higher in urban centres, where vertical farming is usually practised. Besides, there is need for more skilled labour, which is costlier.
- ***Dependency on technology***: Vertical farming is extremely dependent on complicated technologies for lighting and maintaining temperature & humidity. It requires high expertise to manage the operation of vertical farms. In absence of technological support, vertical farming may not be sustainable.
- ***High energy requirement***: Vertical farms require more of energy for provision of supplementary light, maintaining temperature and adjusting humidity.
- ***No standard economics***: The financial benefit of this farming method is uncertain. Under the changing financial situation, the economics of vertical farming needs to be standardized for attracting urban people for the venture.

Types of vertical farming: There are various types of vertical gardening depending on the structure and management options.

Building-based farms: Abandoned buildings are often reused for vertical farming. However, new buildings are also constructed to accommodate vertical farming systems. It becomes easy to manage the plants under direct supervision of the owner.

Shipping-container vertical farms : Unused shipping containers are used for growing plants under vertical farming. The shipping containers serve as standardized, modular chambers for growing a variety of plants. Container-based farms can provide fresh produce year-round at a lower cost.

Deep farms: Deep farm is a vertical farm built from renovated underground tunnels or abandoned mining channels. As the temperature and humidity in the underground are moderate, deep farms require less energy for heating. Deep farms can also use nearby groundwater for growing crops. It requires considerably low costs for establishment of deep farm.

Green Walls

Now-a-days, many people in urban areas are interested for growing plants to enhance aesthetic value and to overcome the ill-effects of pollution. Some have developed a hobby of growing plants to beautify their homes, while plants are grown in commercial and office premises to enhance the beauty of the buildings. In urban areas, inadequate space restricts growing plants in the domestic and commercial premises. It limits the spread of vegetation and greenery in the vicinity, which can be compensated by development of green walls. Green walls can accommodate a large number of plants in a relatively small space thereby providing maximum benefit from limited area. Green walls are mostly developed in individual houses, commercial premises, airports, restaurants, hotels and other establishments of tourist interest. Various green walls have been installed at institutional buildings and public places, which are attracting attention of the visitors. The flexibility of green walls makes them perfect for all types of environment.

Green walls are otherwise known as living walls, vertical gardens, eco-walls or plant walls. Living walls can be fitted to both new buildings and refurbishments. It is a vertical structure used for growing ornamental plants with an intention of enhancing the aesthetic value of the surroundings and augmenting the air quality. Green walls can be erected indoors or outdoors depending on availability of vertical structures. These can be attached to existing host walls or erected structures specifically developed for the purpose. Living walls are particularly suitable for cities due to appropriate use of available vertical surface. These walls are comprised of plants that are inserted into a growing medium and then placed on the wall of buildings to provide greenery using minimum horizontal space.

Advantages

The visual beauty of green wall livens up the area in which it is developed. These attractive pieces make a big impact by decorating the urban environment. If offers alternative for urban gardening and indoor decor. Living walls can make people happier due to its appealing views. Exterior living walls give buildings protection by diverting water away from walls during heavy rain and providing protection from UV radiation.

Green walls help to reduce aggression and anxiety in human mind in urban areas due to the flavour of natural beauty in the immediate environment. Bringing natural elements to the urban areas make the people psychologically active and cheerful. Additionally, the connection with nature through green walls minimizes the stress and mental fatigue.

Green walls help to reduce the temperature of a building due to transpiration and absorption of considerable fraction of solar radiation by the vegetation. Hence, vegetation in green walls can help in mitigation of the heat island effect in the urban environment. Green walls insulate the building and maintain the temperature of building round the year. This can help in cleaning the air from various volatile organic compounds released during construction activities to curb the ill effects of chemical paints and other undesired materials. It helps in cleaning environment by reducing carbon monoxide in the air and filtering out pollutants. The living green walls can reduce levels of nitrogen dioxide and particulate matter substantially. Green walls can also purify the air from dusts and other substances, which can cause asthma and allergies. In urban areas, green walls provide acoustic protection through absorption of sound.

Living walls facilitate water reuse by utilizing the domestic waste water in urban areas. Green walls are suitable for arid areas as the rate of evaporation is slower on vertical walls when compared with horizontal gardens. Green walls are used in the indoors for corporate buildings and offices to create a unique decorative environment. Green walls add colour to the lobby, reception and other places.

Disadvantages

However, there are several disadvantages of green wall including high initial investment, cumbersome installation process, difficulty in management during aberrant climatic conditions and requirement of high-tech management practices. Involvement of sophisticated accessories and high-tech process for construction of green wall necessitates higher initial investment. Most of the green walls are provided with a drip-irrigation system, which also requires high initial investment. The root growth of the plants adversely affects the buildings and other infrastructures. The plant roots can cause damage to the brick and wood by penetrating in the walls. Removal of these outgrowths from the buildings is a costly and labour intensive process. The moist environment on the buildings encourages growth of microorganism that leads to cracks and fissures on the wall.

In some cases, the panels and the growth media are to be replaced intermittently, may be after five to ten years. But in case of tray system, the plants are to be replaced every year. There is risk of damage of the green walls in case of cyclone, earth quake or hailstorm. It becomes difficult to control insect and disease infestation of plants placed at higher positions. The gravitational force limits growing of heavier plants at greater height. Uninterrupted power supply is needed for timely irrigation and maintenance of appropriate light intensity. The chemicals present in fertilizers and agro-chemicals may adversely affect the structure of the wall.

Choice of plant

There are specific plants, which can be suitably grown in the green walls. The selection of plants depends on exposure of the wall to solar radiation, availability of moisture, location, local climate, objective of growing, etc. Plant species may be chosen based on the height of the wall. Ornamental, edible, foliage, flowering and other plants can be selected for growing on the vertical walls. Plants to be grown in vertical gardens should be dense and compact. Ornamental plants add to the aesthetic beauty of the place and make the space eco-friendly.

Some of the foliage plants suitable for vertical gardening are sword fern, money plant, English ivy, etc. Sword fern is a foliage plant most commonly used in vertical garden. Usually, this plant is grown in hanging baskets on green walls. Various fern varieties can survive under low light and some others in medium to bright light intensity. This fern helps in purifying the air. Money plant is a popular vine grown in many households, which is easy to maintain. This plant grows quickly and spreads easily over the wall. English ivy is a hardy and evergreen vine, which can be taken up on the vertical walls. It is attached to the wall and other surface by its aerial root system. This plant has luxurious foliage, which forms a cool green cover on the walls.

Besides, English lavender is another evergreen herb that thrives under hot and dry climate with medicinal and culinary uses. *Aeschynanthus* (Lipstick plant) is suitable for vertical garden due to its simultaneous downward and upward growth habit. It can survive under low light intensity and produce beautiful red flowers. *Davallia fejeensis* (Rabbit's foot fern) is an attractive fern that can be grown with less moisture and light. *Pteris* fern (Cretan brake fern) is suitable for growing under wet condition. This fern can be grown in the lower part of a vertical garden where the water accumulates. Some species of *Dracaena* can be grown on green walls due to low maintenance and minimum requirement of light & water. The leaves are usually bicolour, variegated and dark green. Besides, other plants can also be taken up depending upon local condition and personal preferences.

Giant lily turf, philodendron, geranium, wedding vine, peace lily, etc. are some of the flowering species suitable for vertical walls. The Giant lily turf having dense spreading strap-like dark green coloured leaves and dense foliage blooms with beautiful flowers. Philodendron species can also be taken up on the green walls and are easy to maintain. Those are suitable for both indoor and outdoor areas. Geranium is a plant with deep red, scarlet, pink, white, purple and other colour flowers, which can be taken up on green walls. The flowers last long and add to the beauty of the wall.

Wedding vine plant having white and waxy flowers can be grown vertically on the green walls. Peace lily is also grown on a vertical wall due to its quick growing habit. This plant can tolerate lower light and low humidity. Star jasmine is an evergreen plant, which can be easily grown on vertical walls. However, star jasmine is comparatively heavier plant and requires additional support or trellis.

Structure and growth media

Green walls are constructed with a skeletal structure to support the plants. These are organized into panel and tray systems or free standing walls. Metal frames attached with buildings require a waterproof layer between the growing media and the wall. The panels are modular units, which fit into the metal framework. These are available in various patterns and grid sizes.

There are several methods of growing plants on green walls. In tray system, pre-grown plants are inserted into the wall. Free standing or attached trellis are the least expensive method to create a green wall. A free standing trelly is a frame anchored to the ground. A minimum gap of 14 to 20 cm is maintained between the walls for circulation of air to reduce the moisture build up on building walls. Free standing walls are most versatile and commonly used in the indoors. In this case, the location of the walls and the plants can be changed easily. Use of cables or coated wire can be taken up as an alternative method to the trellis. Cables are anchored with the ground and are attached to roof or wall. Plants grow on the soil and climb up with the support of wire. In this case, the plants require minimum maintenance as compared to other systems of green wall. The spacing of the wires is maintained according to the density of foliage.

Various types of growth media such as loose media, mat media and structural media are used for raising plants on the green walls. A metal framework is installed at little gap from the walls of the building and supports the panel-type planting. The growth media consisting of soil, stone, water or their combination are used for growing plants. In some cases, pre-grown plants are arranged on the green walls. The soil medium is a porous spongy sheet with good water holding and drainage capacity. The soil is packed into a shelf or bag and placed on the wall. Loose soil systems are to be avoided in areas with frequent seismic activity and areas vulnerable to cyclone or heavy rainfall. The height of the structure should be within 2.5 m to protect the loose medium from heavy rain or wind. Adequate care should be taken to prevent detachment of the plants from the wall. Loose-soil medium is best suited for the home gardens, to grow seasonal plants. The media are usually replaced at least once a year in outdoor walls and every two years in indoor units.

Coir fibre mats or synthetic mats are used for growing plants in mat media system. In this method, double layered mats of synthetic fabric with pockets are used. These pockets are filled with growing media for plant growth. The fabric walls are supported on a modular framework layered with a waterproof membrane to protect the building walls. Water and plant nutrients are supplied through drip irrigation system at the top of the wall. Mat media are very thin and cannot support the root systems of full grown plants for more than three to five years. The over growth of roots on the mat prevents water movement. The mats are to be replaced after certain time interval with new ones. These systems are best used on the interior walls of the building and are suitable for areas with low seismic activity. Usually, smaller plants are preferred on mat medium due to their light weight. The mat system is inefficient in water use and requires continuous irrigation.

Structural media are blocks of growth medium of various size, shape and thickness. This medium is most suitable for both indoor and outdoor walls. These can withstand high-winds and seismic activities. Structural media are superior to other media as it lasts long with better performance. These media are durable and can be used for 10 to 15 years.

Care and maintenance

Regular maintenance is necessary for a clean and beautiful living wall. Pruning dead plants and removal of weeds at certain intervals are needed to keep the wall healthy and beautiful. The dried leaves and unwanted plants need to be removed to enhance the aesthetic value of the wall. Keeping the walls clean also helps to avoid infestation of disease and insect. Healthy and disease free planting materials are to be selected for establishment of a healthy green wall. Evergreen perennial foliage plants keep the wall green round the year.

Green walls need frequent irrigation to sustain plant growth. Hence, uninterrupted water supply is must for maintaining a vibrant green wall. Irrigation is required immediately after planting to maintain the moisture level at the root zone. The irrigation system should be designed to use minimum water. This can be achieved through automated or remotely controlled irrigation system, preferably through drip irrigation. The quantity of water delivered is decided depending on the location of the place, season of growing and exposure of the wall to sun light. The water requirement is less for indoor walls as compared to the outdoor walls. Care should be taken to minimize loss of water through run-off. Frequency of irrigation is more during summer months than in winter. The plants can be supplied with nutrients through irrigation water in the form of fertigation.

Green walls can be a boom to enhance the aesthetical value of the buildings, commercial premises and important thoroughfares. Presence of greenery amid the concrete structure of urban dwellings brings mental peace and inner satisfaction of the citizens.

Kitchen Garden

The kitchen garden is a small garden developed near the house for growing vegetables, greens, fruits and medicinal plants to meet the daily requirement of a family. Kitchen garden ensures an inexpensive and regular availability of fresh green groceries to support family nutrition. With rapid growth of cities and urban population, there is need to develop kitchen garden for producing fresh vegetables and fruits. The produces obtained from the garden are healthy, free from agro-chemicals and adulteration. Kitchen gardens can be developed in backyards or terraces or balconies, depending on availability of space. However, the space selected for growing seasonal fruits and vegetables as well as perennials requires proper sunlight. The location of kitchen garden, thus should get adequate sunlight. The area should be well protected from heavy wind. The soil of the sight should be fertile and rich in organic matter.

Backyard of a house is the most preferred area to develop kitchen gardens because of easy accessibility, availability of sunlight and water. The household wastes can be used for meeting the nutritional requirement of the garden. In absence of space in the immediate vicinity of the house, particularly in high rising buildings, kitchen garden can also be planned on terraces or roof tops with plants to be grown in pots and containers. The size of the garden and selection of the plants are decided as per availability of space. Even in terraces or roof tops, number of plants can be accommodated in layers of plants on the walls or frame structure.

Advantages of kitchen gardening

- Availability of home grown seasonal vegetables and herbs free from harmful pesticides.
- Kitchen gardening involves the family members in physical activity.
- Presence of plants refresh the air around
- Provides opportunity for recycling of household wastes and its efficient use as compost
- Producing greens, vegetables and herbs in kitchen garden reduces the overall expenses of the household
- It is educative for children and they know about plant growth cycles
- The surplus produces can be shared with neighbours and friends, which gives a pleasant feeling

Planning for vegetable crops

It is required to prepare a proper planning for growing vegetable crops in the kitchen garden. The following aspects are to be considered while designing a kitchen garden.

- Selection of the vegetables should be based on family requirements. Selection of crop should be as per preference of family members.
- The crop planning should be as per availability of land, family labour, sunlight and water.
- Crop selection depends on time devoted for kitchen gardening and investment capacity of the family. High value crops need more care and resource.
- Inclusion of off season vegetable to minimize expenditure on green grocery of the family.
- Selection of vegetable depends on the knowledge of the grower for easiness in crop management.

- Availability of seeds, seedlings and other materials is also another deciding factor for selection of crops.
- Vegetables like radish, cabbage, cauliflower, tomato, brinjal, etc. can be placed in sunnier part of the garden.
- If there is a shady patch, it can also be utilized to grow turmeric, ginger, pineapple and black pepper to meet the household needs.
- Perennials like papaya, banana, lemon, drumstick, etc. can be planted near the boundary.
- Creepers like gourds, cucumber, yam, etc. can be taken near the fence for best utilization of space.
- Few plants of chili and curry leaf can be planted in a corner to get regular harvest.

At the beginning, the field is divided into small plots depending on the area. The plots/beds should be of convenient size for ease of managing the crop. There should be access pathways between the beds to facilitate operations like weeding, planting, manuring, irrigation, plant protection and harvesting.

Based on above criteria, the crops are decided season-wise as per family requirement and local climatic situation. Care should be taken to obtain diversified vegetables for family consumption in one season. Vegetables of same group should not be grown in the same plot continuously. Growing brinjal, tomato and capsicum in the same plot season after season is to be avoided. An example of kitchen garden lay out with ten plots is given below.

Plot	Kharif	Rabi	Summer
1	Brinjal	Radish	Okra
2	Cowpea	Cabbage	Ridge gourd
3	Cucumber	Tomato	Guanr
4	Pumpkin	Potato	Amaranth
5	Okra	Cauliflower	Bitter gourd
6	Ridge gourd	Capsicum	Poi
7	Colocassia	Broccoli	Pumpkin
8	Bottle gourd	Carrot	Cowpea
9	Bitter gourd	Pea	Brinjal
10	Amaranth	Onion	Bottle gourd

Obtaining best value from kitchen garden

- Growing high-value crops in minimum space. Growing capsicum, broccoli, yam etc. will ensure supply of these costly vegetables from the garden.
- Tomato and water melon need high nutrition, maintenance and water, but produces lots of fruits.
- Beans, cowpea, *guanr* can be grown with less quantity of nutrient supply and care, but produces more.
- Cucurbits like pumpkin, ridge gourd, bitter gourd, cucumber, etc. give ample of production with minimum care.
- Home grown salad leaves are a cheaper alternative to those purchased from the market.

Fencing: Normally, in urban areas the homes are secured with concrete boundaries. In absence of secured boundary walls, kitchen garden can be protected with barbed wire fencing or hedge fencing of *duranta, lantana* or *ipomoea*. The hedge plants are to be closely planted so that stray animals cannot enter in to the garden. It requires pruning to maintain the desired height and good shape. Depending on availability of space, plants like custard apple, *tagara*, yellow oleander (*kaniara*), calotropis (*arakha*) and china rose can also be taken up as hedge plant.

Land preparation**:** Soil forms the foundation of the kitchen garden. The sites for kitchen garden have to be made free of weeds and other unwanted growths. The land has to be tilled properly by spade or trench hoe to pulverise the soil and breaking the clods. Power tiller can also be used for cultivation. The land has to be levelled properly to facilitate irrigation water movement and draining out excess water. Adequate quantity of organic manure is to be applied to the soil before sowing/planting to ensure nutrient supply. In case the soil is rocky or has shallow depth, soil from outside place is to be used. Soil mixture can be prepared by taking compost, cocopeat and garden soil in equal proportions. Leaf mould, vermi-compost, neem oilcake can also be incorporated to fortify the soil. The land has to be divided into small plots as per crop planning. During kharif season, raised beds are to be prepared for ease of planting and drainage of excess rain water as most of the vegetables and greens are susceptible to water stagnation. During rabi and summer seasons, the crops are planted in levelled beds with provision of irrigation channels.

Selection of seed**:** Care should be taken to select good quality seeds of recommended variety for growing in the kitchen garden. The seed should be healthy, pure and free from diseases. Obtaining seeds from a dependable source will ensure good quality. Certified seeds procured from reliable sources result in higher production. If less number of plants are to be transplanted, it is advisable to purchase seedlings from nearby nursery.

***Soil solarization*:** Soil solarization can be done to disinfect the soil of kithen garden. Soil solarization is a simple method to control insect, pathogen and weeds in the soil by using solar radiation. The process involves covering the ground with transparent polyethylene cover to trap solar energy. Temperature of the soil is increased to kill bacteria, fungi, insects, nematodes, mites, weeds, etc. For solarisation, the soil is to be tilled and weeds & debris are to be removed. Watering is done to wet the soil completely. The area is covered with transparent polythene sheet. Use of black polythene is to be avoided for the purpose. The polythene is to be covered with clods or small bricks so that it is not disturbed by wind. The polythene is to be kept on the soil for about 3-4 weeks during summer season. Then the polythene is removed carefully. The pathogen, insect and weed seeds get damaged by this process.

***Seedling raising* :** Kitchen garden requires few number of seedlings of different vegetables, which can be purchased from market or raised by the garden owner. Generally, seedlings of brinjal, tomato, chili, capsicum, cauliflower, cabbage and broccoli are used in the kitchen garden. These can be raised in a small nursery bed prepared in one corner of the kitchen garden. A raised bed of 15 cm height, one-meter breadth and convenient length is prepared. The soil of the bed should be free of clods, weeds and debris. Well decomposed organic manure or vermi-compost and neem oilcake are mixed in the soil. Seeds of desired vegetable crops are to be put in lines and covered with light soil, compost or straw. The bed is watered by rose can. Care should be taken to control insect and disease attack. Seedlings are uprooted at 3-4 weeks after sowing depending on the crop and growing season.

Seedlings can also be raised in pro trays. Seedlings of tomato, chili, capsicum, cauliflower and brinjal can be raised successfully in pro trays. The cups in the tray are filled with growth medium prepared by mixing coco peat, compost, soil and sand. The coco peat is a bi-product of coir industry and it has high water holding capacity. One seed is sown in each cup and covered with growth medium. After sowing of seeds, the trays are stacked one over another for 3 to 6 days depending on the crops. The entire stack is covered with black polyethylene sheet till the seeds germinate. The stacked trays are spread once the seeds germinate. The trays are watered lightly every day depending upon the prevailing weather condition. The seedlings are hardened by withholding irrigation before transplanting. The seedlings are ready in about 21-30 days for transplanting in the main field.

***Nutrient management*:** Feeding the plants plays a vital role in obtaining good yield. Nutrition management can be done through application of organic matter only or in combination with chemical fertilizer. Many persons prefer growing vegetables organically to get quality produce for family consumption. Organic matters such as farm yard manure, vermi-compost, vermi-wash, household compost, bio-fertilizer and oilcake can be used for the purpose. All the organic matter should be properly decomposed for availability to the plant. Effort should be made to prepare compost from the household organic wastes like fruit peels, vegetable peels, used tea leaves, leftover food, debris of plants, weeds, etc. These are to be decomposed in a separate pit prepared in a corner of the garden. Addition of fresh cow dung or cow urine in the pit will accelerate the process of decomposition. The pit is to be filled with organic matter and covered with mixture of soil and cow dung. After 3-4 months, the compost become ready and can be successfully used in the garden. This will help in keeping the premises of the house clean and reduce the expenditure on purchase of organic matter or chemical fertilizer for the kitchen garden. Additionally, *Azospirillum, Azotobacter*, phosphobactor, etc. can be used for improving soil fertility.

***Staking*:** The creepers and trailing plants are to be provided with support for climbing up. Staking can be done using branches of trees, bamboo/wooden sticks, plastic ropes, etc. Staking prevents the fruits from touching the ground and enhances yield. Trellis is made by using wires, ropes, sticks, etc. to facilitate growth of cucurbits. Arches and pergolas can be used to support climbing plants to accommodate more crops at ground level.

***Weed management*:** The kitchen garden is kept free from weeds by hand weeding at regular intervals. It is a means of physical exercise and gives the garden a clean look. Small garden tools like garden hand hoe, spade, trench hoe, garden rakes, etc. can also be used to remove weeds from kitchen garden. If possible, application of organic mulches like straw, compost, grass clippings, newspaper, dried leaves, etc. in between the rows of plants can suppress weed growth. The bunds and irrigation channels are to be kept weed free.

***Irrigation*:** The vegetables, flowers and herbs grown in kitchen garden require regular watering for better growth. Vegetable crops need thorough watering to keep the soil moist. In soils with low water retention capacity, frequency of irrigation is to be enhanced. Mulching vegetable beds with leaves, manure or compost also improves water retention capacity, thereby reducing the irrigation frequency. There should be adequate drainage facility as most of the garden crops do not withstand water stagnation. Kitchen waste water from the house can also be diverted to the garden for irrigating the plants.

Plant protection: In general, chemical methods of pest control measures are to be avoided in home gardens. Preventive measures are to be adopted to check disease and pest infestation. Use of resistant crop varieties, clean cultivation, soil solarisation, seed treatment, soil application of *neem* or *karanja* oilcake, adoption of suitable crop rotation, growing trap crops, etc. are some of the measures to prevent spread of insects and diseases. Application of *neem* based pesticides, *neem* oil, use of light trap, yellow sticky trap and pheromone trap can control pest damage.

Harvest: Harvesting should be done at proper stage of maturity and depending on the purpose of use. Harvesting is to be done in such a manner to get the produce regularly. Plucking of fruits or pods can give continuous harvest. It will help the family to get different types of vegetables daily. In case of cucurbits, harvesting of one fruit will facilitate for production of new fruits. Indeterminate tomato varieties will provide continuous yield for a month or more. Cutting of greens will allow the existing plant to regenerate for additional production.

Nutritive value of vegetables

The vegetables are very good source of fiber, protein and carbohydrate. The average nutritive values of some of the vegetables are given below. However, this may vary depending on variety, growing season, management practices and local climatic condition.

Percentage of nutrient content in vegetables

Vegetable	Carbohydrate	Protein	Fat	Fiber
Beet root	10.0	1.3-1.8	0.7	0.6-3.1
Broccoli	5.3	3.1-4.0	0.3	1.3-3.0
Brinjal	6.7	0.7-2.4	0.7	0.9-2.5
Cabbage	5.4	1.4-3.3	0.7	0.6-3.4
Carrot	10.1	0.6-2.0	0.7	0.6-2.9
Cauliflower	5.2	1.8-3.4	0.3	0.8-2.4
Colocassia	34.6	0.5-2.9	0.1-0.5	0.4-5.1
Cucumber	2.8	0.6-1.4	0.2	0.3-0.7
Onion	8.6	0.9-2.2	0.35	0.5-1.7
Pea	15.6	4.6-8.2	0.8	1.8-5.5
Pumpkin	4.9	0.6-1.8	0.2	0.5-1.3
Potato	25.2	1.6-2.3	0.1	0.3-2.4
Radish	3.6	0.7-1.2	1.1	0.5-1.0
Spinach	3.5	2.3-5.1	0.3	0.6-2.7
Tomato	4.7	0.7-1.2	1.26	0.4-1.8
Yam	27.6	1.5-2.4	2.2	2.1

Yearly Planning of Kitchen Garden

The month-wise planning of various activities in the kitchen garden helps to get continuous supply of fresh vegetables round the year. A tentative planning is given below. However, harvesting is a continuous process and is it done every week depending on the crop.

May

- Cleaning of land and land preparation for sowing
- Layout and plotting of the garden
- Application of organic manure
- Soil solarization
- Procurement of seeds to be grown during rainy season

June

- Seed treatment
- Seed sowing after initial irrigation
- Application of manure and/or basal fertilizer
- Raising of seedlings

July

- Transplanting of seedlings
- Staking of creepers and cucurbits
- Weeding
- Topdressing of manure/fertilizer
- Provision of drainage

August

- Adoption of plant protection technique
- Weeding and hoeing
- Top dressing of manure and fertilizer

September

- Procurement of seeds to be taken up during winter season
- Land preparation for winter vegetables
- Sowing of pre-rabi crops

October

- Application of organic manures/oil cake/bio-fertilizers
- Seed treatment and sowing
- Raising seedlings for winter season
- Application of basal fertilizer

November

- Transplanting of seedlings
- Topdressing
- Irrigation
- Weeding

December

- Weeding and hoeing
- Second top dressing
- Irrigation
- Management of disease & pest

January

- Procurement of seeds to be taken up during summer season
- Land preparation for summer vegetables
- Application of organic manures/oil cake/bio-fertilizers
- Seed treatment and sowing
- Mulching
- Raising seedlings for summer crop

February

- Transplanting of seedlings
- Top dressing
- Irrigation
- Staking of creepers
- Weeding, hoeing and earthing up

March

- Plant protection
- Irrigation
- Weeding and hoeing
- Top dressing of manure and fertilizer

April

- Cleaning of the plots
- Irrigation
- Plant protection measures

3

Flowering Plants

Floriculture is the most important aspect of urban agriculture, which fulfils the need for beautification, recreation, entertainment and income earning. In broader sense, it includes growing flowering and ornamental plants in gardens, rooftops and terraces. Floriculture comprises of house plants, flowering plants, pot plants, cut flowers and decorative foliage. It also deals with cultivation of cacti, orchids, succulents, bonsai, ferns and plants grown in landscape gardening.

Role of flowers in our life

- It is a unique gift of nature to relax human mind
- The fragrance and beauty of flowers bring joy and happiness
- Flowers are used as a symbol of grace and sophistication
- Flowers are used in religious rituals for offering during worship
- Flowers are essentially needed during festive occasions, marriages, birthdays, social functions, official meetings, etc.
- Well-maintained garden reflects the aesthetic value of the house
- Commercially, many flowers are used in preparation of perfumes, pharmaceuticals and beauty care products.

Flowering plants based on growth habit

Flowering plants are of different types depending on morphology, life cycle and growing season. A brief outline of classification of various flowering plants is discussed here.

Annuals: Annual flowering plants complete the life cycle in one growing season. The duration of the cycle may be of several months depending on the species, season and place of cultivation. These flowering plants are useful in landscaping due to their short life span, which creates avenue for accommodating more plants in different seasons of a year. These flowers are more popular in urban households for attractive look, easy maintenance and early production. Some of the commonly grown annual flowering plants are marigold, chrysanthemum, petunia, aster, zinnia, coxcomb, etc.

Biennials: These plants complete their life cycle in two growing seasons. The vegetative growth takes place during the first season and production of flowers & fruits is done during the second season. These plants can be taken up in the backyard or on rooftop/terrace. Example of biennial flowers are sweet william, hollyhock, canterbury bells, forget-me-nots, foxglove, honesty, etc.

Perennials: Perennial flowering plants are mostly woody species with a life span of more than two years. These plants live for several years with vegetative and reproductive growth. Some of the perennial flowers are rose, bougainvillea, hibiscus, ixora, adenium, magnolia, clitoria, golden trumpet, vinca, etc. These plants ensure continuous supply of flowers to meet the daily requirement of urban families, particularly for daily offering to God. Most of these plants need minimum maintenance. Some of these species are suitable for green hedges.

Flowering plants based on use

Flowering plants may also be categorised as per their use. Some plants are grown indoors, while some others are grown in the outdoors. Depending on use, classification of the flowering plants is depicted below.

Cut flower: Some flowers like rose, anthurium, tuberose, carnation, orchid, chrysanthemum, lily, gerbera, etc. are used as cut flowers. The flowers are cut at blossom or bud stage along with stems, branches and leaves. Cut flowers are used for preparation of bouquets or for decorations. Thus, these flowers are in high demand in urban and peri-urban markets. Due to high market value, these flowers can be grown commercially for income generation.

Potted plants: Potted plants are preferred by urban people as those can be accommodated in comparatively less area and can be conveniently kept on rooftops/balconies/terraces or in the living rooms. These plants are widely used for beautification of commercial establishments. Potted plants are suitable, both for indoor and outdoor conditions in the house and in small gardens. These are usually annual flowers like geranium, aglonema, starflower, petunia, etc.

Bedding plants: Flower beds in front of the residential house, public office or commercial establishment improves aesthetic value of the building. Generally, annual flowering plants like marigold, chrysanthemum, petunia, aster, zinnia, gerbera, salvia, pansy, etc. are grown in beds. Flower beds can also add to the income of the grower, if taken up commercially.

Hanging plants: In this condition, the plants are grown in the basket, pot or other containers and hung in the balcony, terrace or other convenient place.

Generally, annual flowering plants like geranium, table rose, petunia, orchids, etc. are planted in hanging containers to add to the beauty of the premises.

Dried flower: In some areas, the flowers are dried and preserved for future use. These flowers are used during festive occasions. Decorative wall hangings can also be prepared from the dried flowers. Now-a-days, some preservatives are used for storage of dried flowers. The bouquets and decorative products prepared out of dried flowers are used in urban households for beautification of the building. The flowers of Ageratum, Lavender, Pansy, Baby's Breath, Globe Amaranth, etc. are used after drying for preparation of garland and other decorative products.

Landscape plants: These are the flowering plants used for beautification of the landscape area. These plants can be grown either in open space or under shade. Sometimes, these plants are taken up to cover up unfertile, marginal and barren land. Some of the flowering plants like vinca, aster, bougainvillea, sedum, hibiscus, geranium, etc. can be taken up for landscape designing.

Rose

Rose is the most beautiful flower grown for improving the aesthetic value of the garden, premises and other places. This flower is grown in the home gardens or fields for commercial purpose. It can be successfully grown in minimum space by use of pots or containers; placed on the veranda, terrace or rooftop. Planting in pots provides opportunity for precise management of the plant and efficient utilization of sunlight, water and nutrients. Growing potted rose is a unique option for the home garden and rooftop in the urban areas.

There are many types of roses, depending on the plant type, growth habit and flower size. Varieties with small plant stature and bushy type are preferred for growing in pots.

i. Hybrid Tea with bushy appearance is one of the most popular type of rose grown in India. The flowers are having 30-50 petals and suitable for use as cut flowers. Some of the popular varieties are Double Delight, Evening Star, Dil-ki-Rani, Happiness, Golden Giant, Kiss of Fire, Sea Pearl, Paradise, etc.

ii. Polyanthas rose requires low maintenance and shows resistance to diseases. This rose provides an eye-catching sight when grown along the hedges or in the pots. The shades of red, pink and white are very attractive for the garden. Some of the examples are Sneezy, Rashmi, Pink Spray and Fairy Rose.

iii. Floribunda rose is a cross between Polyantha rose and Hybrid Tea. The plant is dense with clusters of big flowers in shades of yellow, white, pink, purple and orange. As a low-growing shrub, it is suitable for hedges. Some of the varieties under floribunda are Mahak, Kusum, Prema, Rajamani, Sindoor, Chandrama, Golden, Summer Snow, etc.

iv. Grandiflora rose is a cross between Floribunda and Hybrid Tea. The plant bears a cluster of flowers on a single stem. It has attractive shades of yellow, orange, red, pink and purple. Some of the varieties are Earth Song, Pink Parfait, etc.

v. Climber rose looks like a vine. Those are having long, stiff canes and can be trailed by using a support. Horizontal trailing stimulates better blossoming. The big flowers are very attractive. Some of the examples are Kanyakumari, Eden and May Queen.

vi. Landscape roses produce flower throughout the year. These plants have a sprawling habit, which makes those great space filler in the garden. The plants require minimum maintenance. Flower Carpet Coral and Flower Carpet Scarlet are two examples of landscape rose.

vii. Shrub rose is having features of traditional and modern roses. Flowers have double petals and are closely attached to each other. Examples of shrub rose are Iceberg and Beach.

viii. Damask rose is a cross between *Rosa moschata* and *Rosa gallica.* The flower colour is deep pink or light pink. This variety is used for preparation of rose oil. Noorjaha, Him Jwala and Him Himroz are some of the examples of Damask rose.

ix. Bourbon rose is a cross between old blush China rose and Damask rose. This variety is used to produce rose oil due to its pleasant fragrance. Flowers are snow white or dark pink in colour. Zephirine Drouhin is an example of Bourbon type rose.

x. Alba rose is one of the oldest varieties of rose. This plant can be grown under shade and exhibits considerable resistance to diseases. The flower colour is pink to white shade with a sweet aroma.

xi. Kashmiri rose is mainly used as cut flowers. The flowers are bright red in colour and having mild scent. The petals are velvety soft and used for bouquet preparation.

xii. Miniature roses are small in size with short and bushy plants. These plants are hardy and bloom continuously for 2-3 weeks. Those are suitable for hanging pots. Some of the varieties are Baby Love, Dazzler, Lavender Jewel, Beauty Scarlet, etc.

Generally, budded plants are used for propagation of rose. Climbers, polyanthas and miniatures are multiplied by stem cutting. While selecting budded plants, care should be taken to avoid plants with wild shoots.

Ideal planting time for rose varies from October to November. However, rose can be planted in other months depending on local climatic condition. The potted roses are to be kept in places to receive at least seven hours of direct sunlight. The pots are to be placed at a distance of two feet to ensure better sunlight and air circulation. Being a perennial plant, it requires regular nutrient supply through application of organic manure and chemical fertilizer. Roses are heavy feeders and need frequent feeding when grown in pots.

Based on the age of plant, 5-6 kg of well decomposed manure may be applied in 3-4 splits to each plant during a year. If required, fertilizer mixture of urea, single super phosphate and muriate of potash may be applied in the ratio of 1:3:2. Foliar spray of urea @ 2g/litre of water is helpful to obtain better plant growth. Micronutrients like boron, zinc, molybdenum, magnesium, etc. can also be applied to get good quality flowers. Now-a-days, several readymade fertilizer products are available in the market for nurturing the rose plants. Application of excess dose of fertilizer can damage the plants. Direct contact of fertilizer with the leaves may have scorching effect on the plant. Immediately after planting, the plant should be watered to help quick establishment. The rose plants are to be watered as and when required so as to keep the root zone moist. Stagnation of water in the pots may result stunted growth and wilting of plants due to root rot or other infection. Care should be taken to avoid watering on the leaves as this may lead to fungal infection.

Rose plants need to be pruned during third week of October to get better flower production. Of course, it varies depending on local climatic condition. Usually, flowering starts 6-7 weeks after pruning. In older plants, the shoots are pruned up to half of the length by keeping five to six eyes on each shoot. Normally, the stems are cut in a slanting manner above the eye. In case of climbing rose, pruning is not required except removal of unhealthy and dead branches. Hybrid tea and floribunda requires hard pruning by keeping three to four shoots in the base. Sometimes, suckers emerge from the rootstock with a characteristics wild plant. These are to be removed for efficient growth of the budded plant.

Rose is a heavy feeder and quickly exhausts the soil in the pot. Hence, the pots used for growing roses, need to be refilled with freshly prepared pot mixture in every two or three years. Moreover, there is accumulation of salt and minerals in the soil due to application of fertilizer, which may damage the plant.

Rose is susceptible to several pests and diseases. Common insects found in rose are aphid, thrips, hairy caterpillar, cut worm, etc. Aphids and thrips are the sucking pests, which suck out juice from the buds and leaves resulting in stunted growth of the plant. These can be controlled by spraying imidachloprid @ 2.5 ml/10 litre of water. Hairy caterpillar and cut worms feed on the flowers and leaves resulting in defoliation of leaves and loss of flowers. Spraying of profenophos @ 2.0 ml/litre water can control these insects. Roses are also infected by fungal diseases like powdery mildew and die back. In case of powdery mildew, there is appearance of white spot on the leaf surface. The symptoms of die back disease are drying of tips of the plant. These diseases can be controlled by spraying propiconazol @ 1.0 ml/litre water. However, several neem based products are available in the market for control of the insects and diseases. Clean cultivation and proper management practices can reduce the risk of disease infestation.

The flowers are to be plucked at the bud or pre-blooming stage, when the colour has fully developed. In some cases, flowers are cut along with certain length of stem, when one or two petals start unfolding. The length of the stem varies depending on grade of the flower. It is better to pluck the flowers either early in the morning or late in the afternoon. The flowers need to be kept in plastic container filled with water and preservatives like silver thio-sulphate to enhance the shelf-life. Flowers can be preserved for longer duration under low temperature. Usually, one rose plant yields nearly 25-30 flowers per season.

Marigold

Marigold is the most commonly grown flower in gardens and rooftops of urban houses. It is largely used in religious and social functions as loose flowers and for preparation of garlands. Commonly grown marigold species are *Tagetes erecta* (African marigold) and *Tagetes patula* (French marigold). African Marigold is a tender, herbaceous annual plant producing large flower heads. It needs staking due to tallness. Some of the varieties of African Marigold are Snowbird, Sugar and spice, Royal yellow, Inca gold, Inca orange, Inca yellow, etc.

French marigold is a compact, bushy annual plant originated from Mexico. The flowers are of different colours such as red, orange and yellow. Various cultivars of French marigold are Golden Boy, Harmony boy, Arka Pari, Scarlet sofia, Arka Honey, Bonanza orange, Red cherry, etc.

Marigold can be suitably grown in deep, fertile and well drained soil with good water holding capacity. The desirable pH range of soil for marigold cultivation is from 7.0 to 7.5. Marigold can be grown throughout the year. It requires mild climate for luxuriant growth and profuse flowering. Optimum temperature requirement for germination of seed is 18^0 C to 30^0 C.

Marigold is propagated through seeds. Seeds should be healthy and free from disease and insect attack. Seedlings are prepared in the nursery bed for planting. Seedlings are grown in nurseries to protect from adverse climatic condition. Seeds are sown during June-July for rainy season crop, September-October for winter season crop and January for summer season crop. Of course, the time of sowing varies depending on local climatic condition. The seedlings are to be properly protected from attack of insect and disease. Uniform growth is required to get good seedlings. Healthy and robust seedlings are desired for obtaining higher flower yield. After 25-30 days of sowing, the seedlings are ready for transplanting in the main field. Healthy and good quality seedlings can also be procured from well maintained nurseries.

Transplanting is done in small flat beds or in shallow furrow. In heavy soil, it is usually transplanted on ridges. Planting the seedlings on ridges helps in better survival and growth during rainy season. Transplanting can be done during the month of July-August for rainy season, mid-October for winter season and February for summer season crop. Normally, planting is done at a spacing of 60-75 cm X 60 cm. This may vary depending on the variety, growing season and management practices.

Marigold plants can also be grown in pots or containers. As African marigold grows up to a height of one meter, it requires large containers for growing. French marigolds are small and bushy. These plants grow up to a height of 15 to 20 cm. Hence, this can be grown in small pots.

Marigold plant can be grown in a pot of 15 cm diameter. Sometimes, 2-3 plants are grown in a pot with 30 cm diameter. The pot or container should have a hole in the bottom for proper drainage. The potting mixture should be of good quality and light weight. Pot mixture can be prepared by mixing equal quantities of garden soil, sand and farm yard manure.

The pots are to be placed suitably so that the plants can get adequate sunlight for at least six hours. The plants are to be watered, when the top 2.5 to 5 cm of the soil is dry. The pots are allowed for alternate wetting and drying for better growth and flowering. Standing water should be avoided in the pots as the wet soil condition triggers root rot disease. Pinching the tips of the plants once or twice is required to get bushy plants and more number of flowers per plant.

Application of farm yard manure, vermi-compost and composted household wastes are suitable for fulfilling the nutrient requirement of marigold plants. However, application of water-soluble fertilizer once in a month can result in better flowering. Too much of nitrogen application may lead to excessive foliage growth with few flowers. Now-a-days, various organic products and plant growth regulators are available in the market to accelerate blooming of the plant.

Various insects and diseases attack marigold plant, which adversely affect plant growth and flower yield. The infestation is more acute during rainy season as compared to other seasons. Thrips, plant hoppers, leaf miners, leaf eating caterpillars, aphids, white flies, etc. are some of the insects affecting the plant. These insects can be controlled by application of neem oil or neem based pesticides. Some of the insecticides like imidachloprid, profenophos, methyl dimeton, etc. can be used to control the insects.

The major diseases affecting marigold are leaf spot, damping off, early blight, powdery mildew, etc. These fungal diseases can be controlled by application of mancozeb, propiconazol, copper oxychloride, etc. Neem oilcake may be mixed with pot mixture to minimize the infestation of disease and insect.

Marigold is harvested about 50-60 days after transplanting. Marigold flowers are harvested when the flowers attain full size. Plucking of the flowers should be done in cool hours of the day. The field should be irrigated before plucking so that flowers can be stored for a comparatively longer period. African marigold yields about two kg flowers per square meter area. In case of French marigold varieties, the yield is nearly one kilogram per square meter area.

Chrysanthemum

Chrysanthemum is an important cut flower with wide range of colours and varying forms. Based on the purpose of use, there are two categories of Chrysanthemum viz. garland purpose and cut flowers. The varieties suitable for garland purpose are Baggi, Basanti, Indira, Rakhi, Red gold, Sharad Mala, Meera, etc. Suitable varieties for cut flowers are Apsara, Arctic, Charlia, Jubilee, Kundan, Purnima, etc.

Chrysanthemum can be grown under outdoor conditions except during extreme temperature and excess rainfall. The optimum day and night temperature requirement for chrysanthemum are 18-21°C and 10-16°C, respectively. Sandy loam soils are preferred for cultivation of chrysanthemum due to optimum aeration. Chrysanthemum thrives well in slightly acidic to neutral soils with pH ranging from 6.5 to 7.0.

Propagation is done by use of cuttings and suckers. Mostly cuttings and suckers are used to obtain true to the type plants. Most of the plants do not set seeds. Rooted suckers are planted in the field during January. Terminal cuttings are taken from healthy stock plants during June and transplanted after rooting. Rooting takes place within 3-4 weeks. Top 2-3 leaves are removed during first pinching at 8-10 leaf stage of the plant. In standard types or garland flowers, axillary buds are removed. In case of cut spray types (cut flower), terminal buds are removed by allowing axillary buds to bloom.

Chrysanthemum requires higher quantity of nitrogen and potassium during vegetative growth. The pot mixture should contain soil, farm yard manure and leaf mould in equal ratio. Little quantity of bone meal or super phosphate is to be added to the pot mixture to enhance plant growth. For quality flower production, well rotten FYM and vermi-compost should be applied. Application of chemical fertilizer in judicious quantity results in higher yield. Bio-fertilizers such as Azotobacter and PSB can be applied by dipping the roots of the cuttings. The plants should be watered twice in a week or depending upon the soil moisture condition.

Staking is required to maintain the shape of the plant and obtaining good quality flowers. Staking is to be provided when the plants are 30 cm tall. In standard cultivars, number of stakes depends upon number of side branches. In spray types, 3-4 stakes are inserted in the pot and are tied with a string to support the canopy and flowers.

Chrysanthemum is affected by several insects like thrips, aphids and leaf eating caterpillars, etc. These can be controlled by spraying of Acetamiprid @ 0.3 g/ litre water or Indoxacarb @ 1 ml/ litre water. Several diseases like leaf spot, root rot, wilting, etc. infect the plants. Application of Copper oxychloride @ 2.5 g/litre or Trifloxystrobin + Tebuconazole @ 0.75 g/litre or Difenoconazole @ 0.5ml/l can control these diseases.

Spray cultivars (cut flower) are ready for harvest when the lower 2-3 flowers are opened, whereas standard types are plucked when a few outer ray florets are unfurled. Plucking can be done during evening or morning hours. On an average, a single plant yields about 0.75 – 2.0 kg loose flowers depending upon the variety.

Jasmine

Jasmine flowers are white in colour with attractive scent. These are used for making garland and floral decoration. The oil extracted from the flower is used for production of perfume. Among various fragrant species; mogra has a high commercial value. Mogra has a short post-harvest life and hence its cultivation is concentrated near the urban centres. The mogra species blooms during summer season.

It requires well drained and fertile soil. Jasmine cannot tolerate waterlogging. It requires hot and dry climate for good yield. Yield of the flower is influenced by temperature and relative humidity. Under high temperature and high humidity, although the vegetative growth is vigorous, the bushes are less productive. There are several types of jasmine such as mogra, Jui, Chameli or Jai and Kurt da. The varieties grown under mogra are Motia, Madanban, Ramban, Khoya, Single Mohora, Double Mohra, Rai Japanese, Virupakshi, etc. The prominent Jui varieties are Co-1 Mullai, Co-2 Mullai, Pari Mullai, Long round, Short round, Long point and Medium point. Some of the varieties under Chameli type are Co-1, Thimmapurum, Coimbatore and Pink pin. Jasmines are generally propagated by cutting, layering, budding and grafting. However, the common method of propagation is through cutting and layering. Rainy season is suitable for plant propagation.

Jasmine is a perennial plant, which remains productive for 10-15 years. Hence, land should be properly prepared with addition of manures. Planting is usually done in rainy season in the pits or pots. The pits or pots should be filled in with a mixture of well decomposed farm yard manure, sand and soil. Well rooted cuttings are planted in the pit or pot. Planting may be done at a spacing of 1.8 meter from plant to plant and 1.8 meter from row to row. Wider spacing may be given for climbing varieties.

Moderate watering is good for jasmine. Irrigation should be given at an interval of 3-4 days till establishment of the plants in absence of rain. Subsequently, irrigation may be provided at an interval of 15 days. After cessation of flowering, watering should be completely stopped till pruning. The plants need pruning to encourage new shoots and enhancement of flowering. In Mogra type, flowers are borne on new growth. Hence, pruning stimulates new growth, results in higher flower production and keeps the plant in manageable size.

The plants are to be provided with 10 kg FYM, 100 gram nitrogen, 150 gram phosphorus and 100 gram potassium per plant per year. Micronutrient mixture of magnesium, zinc and boron may be applied to enhance the flower production. The fertilizers may be applied in 2-3 splits in a year. The potted plants can be manured with FYM, compost or vermicompost.

Jasmine plant is affected by insects like whitefly, thrips, leaf webber, etc. These can be controlled by application of profenophos, thiomethoxam, imidachloprid, etc. The common diseases of jasmine are blight, rust and wilt.

The symptoms are development of necrotic areas & discoloured patches on leaves and dropping of leaves. The diseases can be controlled by application of copper oxychloride, propiconazole and plantomycin. The insect and disease can also be controlled by spraying of neem oil or neem based pesticides.

Jasmines start flowering from first year of planting. Jui and Mogra flowers are harvested from March to June. Chameli flowers are harvested round the year. Mogra is harvested when the flower buds are fully developed and unopened. It is plucked in the early morning. For preparation of garland, fully developed unopened buds are picked. But for extraction of oil, flowers are harvested after complete opening.

The yield varies depending upon the type of jasmine and management practices. However, mogra type yields about 0.8 to 1.0 kg flower per m^2 area. The yield from Chameli type plant is slightly higher than the mogra type plant.

Dahlia

Dahlia is one of the most attractive flowering plants, which is normally grown during winter season. It can also be grown throughout the year in cooler climates at higher altitude. This is a popular flower with diversified colour, shape and size. Profuse flowering and easy maintenance have made it popular among the urban gardeners. Dahlia occupies conspicuous place in any garden and extensively used for exhibition, garden display and home decoration. This can be grown both in the bed and in the pot or container. Dahlia species with large flowers and bright colours add to the beauty of the terrace or rooftop garden.

Dahlia grows well in properly drained and fertile soil. Organic matter rich fertile loamy soil is suitable for dahlia flower production. The ideal pH requirement is around 6.5. The plant requires bright sunshine, cool atmosphere and frost free condition. It is generally planted during September-October in the plains and April at higher altitudes. Dahlia cannot withstand extreme cold or hot climate. Moderate temperature with mild humidity is suitable for cultivation.

Dahlia has several varieties, which are categorised under different types. The varieties under giant decorative type are African queen and Alvas Supreme. The large decorative varieties are Alden Galaxy and Arthur Hambly. Varieties under medium decorative type are April Dawn and White Blend. Cristine Hammett is a type of miniature decorative plant. Some other popular varieties of dahlia are Croydon, Masterpiece, Fancy Free, Kelvin Rose, Arab Queen, etc.

Dahlia is propagated by seed, tuberous roots and cuttings. In case of tuberous roots, whole root or root cuttings are planted. Dahlia can be grown either in pots or on bed. Staking is required to protect the plants from strong wind. Proper nutrition may be done to obtain good quality flowers. FYM may be mixed with the soil @ 5 kg per m^2 before planting. Fertilizer may be applied @ 10 gram N, 12.5 gram P_2O_5 and 10 gram K_2O per m^2 area. However, fertilizer application may vary depending upon the soil status and variety grown. Planting is done at a spacing of 75 cm for tall varieties, 60 cm for medium varieties and 30–45 cm for dwarf varieties. Shading of 3-4 days is required for establishment of the cuttings. The lateral shoots arising from the tuber (except the strongest one) are to be removed. Pinching is required at initiation of two or three pairs of leaves to induce branching.

Dahlia plant is affected by various insects such as aphids, thrips, mites and leaf hoppers. These insects adversely affect the plant growth by sucking the sap. Application of systemic insecticides such as profenophos, imidachloprid, etc. can control the insects. Various diseases like stem rot, wilt, powdery mildew, leaf spot, etc. cause damage to the plants. The plant growth is seriously affected due to spread of the diseases. Application of carbendazim, mancozeb, copper oxychloride, etc. can inhibit the spread of these diseases. Neem oilcake may be added with the pot mixture to check spread of disease. Various types of neem based pesticides can also be used to control disease and insect of dahlia plant.

Dahlia flowers are cut when those are two-third open. The flowers are harvested in the early morning. Just after harvesting, the cut flowers are emerged in water up to the neck of the flower stalk. It should be placed in a dark and cold room for long vase life.

Petunia

Petunia is one of the most popular flowers in urban gardens due to its longer flowering period and colourful blooms. The flowers are of several colours and patterns. It blooms from spring season till late summer. These colourful annual flowers add beauty to the gardens and are often planted in beds, containers and hanging pots. The plants grow up to a height of 15 to 45 cm and spreads on ground area of 45-90 cm. It is a heavy blooming plant with multi variety flowers. It is easy to maintain petunia plant. Depending on flower size, petunias are divided into different groups.

Multiflora: This variety of petunia flowers for longer period. The plants bear more number of smaller flowers and are suitable for summer planting. These are more tolerant to wet weather condition. Carpet series, Primetime series and Heavenly Lavender are popular varieties of multiflora.

Grandiflora: Grandiflora petunia bear larger size flowers and are best suited for growing in containers or hanging baskets. These are highly susceptible to damage by rain. These large flowers are prone to rot during hot and humid summer season. Sugar Daddy (Petunia Daddy Series) and Rose Star (Petunia Ultra Series) are widely cultivated grandiflora varieties of petunia.

Floribundas: The flowers of Floribundas are of intermediate size between the grandiflora and the multiflora groups. Those are free-flowering like the multiflora varieties and produce medium-size blooms. Celebrity series, Madness series and Double Madness series are popular floribunda cultivars.

Millifloras: Milliflora petunias are the smallest flowers as compared to other species. However, flowering continues for a longer period with prolific blooming. Fantasy is the popular milliflora cultivar.

Spreading or trailing petunias: These plants are of smaller height and can spread on ground area of 1.2 to 1.6 meter. The plants form a colourful ground cover as the flowers bloom along the entire length of each branch. These varieties can be used in hanging baskets or pots kept on the rooftop. Purple Wave is a cultivar of spreading variety, which produces large purple flowers.

Petunia can be grown in both hot and cold climate depending on the variety. Grandiflora variety can be grown in India during winter season. Petunia grows well under bright sunshine. It does not perform well in shade. Petunia can be grown in different types of soil with good drainage facility. Soil should be fertile to facilitate desired growth.

Petunia can either be planted in the pot or in the ground. Earthen, cemented or plastic pots of various size with shallow depth are suitable for growing petunia. The plants with attractive flowers of various colour combinations have attractive look in hanging pots. The pots should have adequate drainage facility with provision of at least three to four drainage holes at the bottom. Waterlogging in the pot can cause root rot disease and can damage the plant. Usually, seedlings are transplanted in the pot to get healthy plants. High quality, disease free seeds can be sown in the pots to raise seedling. The pots are to be filled with proper pot mixture consisting of farm yard manure, soil and sand for raising seedling. Seedlings emerge after 10 days of sowing. Seedlings of 4 to 5 leaf stage are ready for transplanting. Generally, petunia blooms after 90 to 100 days of germination.

It is required to grow the plants with proper potting mixture prepared by mixing 40% normal garden soil, 30% compost/vermicompost, 20% sand and 10% cocopeat. Addition of handful of neem oilcake with the pot mixture can prevent infestation of diseases and pests. Regular watering at weekly interval and exposure to sunlight promote multiple blooms. Application of balanced water soluble fertilizer is required to obtain profuse blooming. Petunias are heavy feeders and require nutrient supply at regular intervals. After one to two weeks of transplanting, application of one to two handful of bulky organic compost is helpful. Along with this, application of one teaspoon bone meal per pot once in a month results in better flowering. Proper sunlight, water supply and fertilization will ensure beautiful blooms for a longer period.

By mid-summer, most of the plants produce flowers at the tip of long & leafless stems. At this time, pruning of the shoots at about half their length is advisable to keep petunia tidy and flowering. This will encourage more branching and more flowers. After pruning, fertilizer and water are applied to the plants to trigger new growth. Faded, old or dead blossoms are to be removed to enhance attractiveness of the plant.

Aphids, slugs and snails damage the petunia plant. Salt water solution can be used to control snails and slugs. Care should be taken to avoid spraying the solution on the plants. Aphids can be controlled by application of insecticides like imidachloprid and dimethoate. Petunia plant is affected by various diseases such as root rot, stem rot, powdery mildew, wilt, etc. The plant growth is affected due to infestation of the disease. These can be controlled by spraying of fungicides like cupper oxychloride, mancozeb, propiconazol, etc.

Tuberose

Tuberose is popularly known as Rajanigandha or Nishigandha. It is a sweet smelling bulbous flowering plant with long lasting flower spikes. Tuberose can be grown in pots and beds. The flowers are used for making garlands, bouquets and floral decorations. Essential oil is also extracted from the flower, which is used for preparation of perfume, soap, cosmetics and medicine. Popularly cultivated tuberose is of four types on the basis of petal arrangement. Those are single, semi-double, double and variegated.

Single types bear pure white flowers with one row of corolla segment. These flowers emanate strong aroma and are extensively used as loose flower. It is also used for extraction of essential oil. Some of the promising varieties are Single Mexican, Kalyani Single, Shringar, Prajwal, Arka Nirantara, Hyderabad Single, Calcutta Single, Phule Rajani, Kahikuchi Single, Pune Single, etc. In the semi-double variety, the flowers bear two or three rows of corolla segments on straight spikes.

The flowers of double variety type of tuberose have more than three rows of corolla segments. The flowers are white in colour with pinkish red tinge. Some of the important varieties are Pearl for Double, Kalyani Double, Hyderabad Double, Calcutta Double, Vaibhav and Suvasini. The variegated types have silvery white or golden yellow streaks on leaves. The examples of variegated varieties are Rajat Rekha and Swarna Rekha. Rajat Rekha variety has silvery white streaks along the middle of the leaf. In Swarna Rekha variety, the flowers are double type with golden yellow steaks along the margins of leaf.

Tuberose can be grown in open sunny location with warm and humid climatic condition. The plant flowers profusely throughout the year under moderate climatic conditions. Temperature above 40°C adversely affects length of the spike and quality of the flowers. Very low temperature and frost damage the plants. Humidity and temperature are the two main factors influencing production of tuberose. Tuberose can be grown on a wide variety of soils ranging from sandy loam to clay loam. The soil should be well drained, friable and rich in organic matter. Tuberose cannot tolerate waterlogging even for a short period. Fertile, loamy and sandy soils with pH of 6.5 to 7.5 is ideal for tuberose cultivation. The plots are to be properly cultivated to make the soil fine tilth. The soil should be exposed to sunlight at least for two weeks to destroy the weeds and other pests.

Tuberose is generally planted in March-April in the plains and April-May in the hills. In some places, planting can also be done during July-August. Staggered planting can be adopted to obtain flowers round the year. Selection of suitable bulbs is important for successful tuberose cultivation. Healthy and disease free bulbs are ideal for planting. The tuberose bulbs can be used for planting at 4-5 weeks after harvesting. Planting of fresh bulbs may result in profuse vegetative growth and poor flowering.

The bulbs remain dormant during the winter months due to extremely low temperature. The dormancy can be broken by dipping the bulbs in 4% Thiourea solution for one hour. About 12 to 15 bulbs are required for planting in one square meter area. Large sized bulbs with two cm or more diameter are suitable for planting. Large size bulbs are cut vertically into 2 to 3 sections.

The planting distance varies with the soil and climatic conditions. For economic returns, bulbs are planted at an optimum spacing of 20-30 cm x 20 cm. The depth of planting varies from 3.0 to 7.0 cm depending upon the diameter of the bulb and the soil type. The bulbs are thoroughly cleaned and treated with carbendazim (0.2%) for 30 minutes before planting.

Tuberose responds well to application of organic manures. FYM may be applied @ 2 kg/ m^2. Besides, application of 10 gram N, 5 gram P_2O_5 and 7 gram K_2O per m^2 is to be done for tuberose production. Half of N and full dose of P_2O_5 & K_2O are to be applied at the time of planting and the remaining N is to be top-dressed 45 days after planting. Foliar application of GA3 @ 50 to 100 ppm thrice at 40, 55 and 60 days after planting is beneficial. It increases production of good quality flowers. It is necessary to irrigate before planting to provide optimum moisture in the soil. Subsequent irrigation is given depending upon the availability of soil moisture.

The insects affecting tuberose are thrips, aphids, bud borer, grass hoppers, weevils, etc. These are controlled by spraying of Profenophos or Metasystox @ 2.0 ml / litre water. Besides, neem oil can be sprayed to control the insects. The diseases affecting tuberose are stem rot, flower bud rot, Sclerotial wilt, Alternaria leaf spot, rust, powdery mildew, etc. These can be managed by application of mancozeb or propiconazole.

Flowering starts at 80 to 100 days after planting. Harvesting is done by cutting the spikes daily as those open. Spikes are harvested preferably in the morning or in the evening hours by clipping with a sharp knife or secateurs.

After harvesting the main crop, the flower stalks are cut to the base. The plots are manured and irrigated to grow the ratoon crop. About 3-4 ratoon crops can be taken up from a single planting. The ratoon crop results in less number of florets. Immediately after harvest, the lower portion of the cut spikes should be immersed in water for increasing the vase life.

Hibiscus

Hibiscus is one of the most commonly used flowers in daily life of Indian household. It is particularly used for worship. It is an impressive hedge plant with trumpet shaped flowers in various colours and sizes. Over 200 species of hibiscus and many more cultivars & hybrids are available. Tropical hibiscus varieties are suitable for warm climate. Irrespective of variety, hibiscus flowers are very attractive. There are four broad types of hibiscus.

Swamp Hibiscus (*Hibiscus coccineus*): It is a hardy plant bearing large pinwheel-like flowers

Confederate Rose (*Hibiscus mutabilis):* The flowers are large and showy with white colour at the early stage and subsequently change the colour.

Cajun Cocktail (*Hibiscus rosasinensis)* **:** It is a tropical cultivar with striking orange and yellow ruffled blooms.

Perfect Storm (*Hibiscus moscheutos)*: A hardy and compact type plant that blooms late in the season.

There are numerous varieties and hybrids of hibiscus available for cultivation. Some of the varieties include China rose, Rock Hibiscus, Flower of an Hour, Blue bird, Rose of Sharon, Giant Rose Mallow, Rose Mallow, etc. This is a perennial herbaceous plant and grows up to a height of four meter. The hibiscus plants require exposure to bright sunshine for growth and flowering. It grows well in loamy soil with neutral or slightly acidic reaction. The flowers are of different colours like white, red, pink, yellow, orange, etc.

Hibiscus is propagated from stem cuttings. Four to six-inch long stem cuttings are used for propagation purpose. The leaves of the stem cutting are removed except the top leaves. Treatment of the stem cutting with rooting hormone facilitates early establishment of the plant. It takes a couple of months for the plants to fully establish. Supply of higher doses of nitrogen and potassium encourages abundant, healthy blooms with good colour. Application of bulky organic manure enhances flower production. Potted hibiscus need repotting in a couple of years.

Hibiscus plants need moist soil condition. Plants need regular watering during growing season. In this case, watering is done when the top one inch of the pot mixture dries out completely. Over watering is harmful for the plant. Pruning is done by removing the dead, damaged and diseased branches of the plant, which helps in better air circulation and keeps the plant healthy. Normally, pruning of the plant is done after flowering. Hibiscus plant is associated with several problems as discussed hereunder.

***Yellow foliage*:** Sometimes, the leaves of the plant turn yellow, which may be due to sudden changes in weather conditions, improper watering or inadequate fertilizer application. However, seasonal change in temperature may also result in yellowing of the leaves.

***Dropping of buds*:** Extreme temperature, inadequate light or humidity and over watering or water deficit can result in bud drop.

Hibiscus plants are affected by various insects such as aphids, white fly, thrips and cutworm. These can be suitably controlled by application of imidachloprid, profenophos or dimethoate. Besides, neem oil and neem based pesticides can also be used to control the insects. Sometimes, leaf spot and root rot diseases are observed in the plants. These diseases can be controlled by application of propiconazol or copper oxychloride.

Crepe Jasmine (*Tagara*)

Crepe jasmine or *Chandni* is an evergreen flowering shrub. It is also called as *Tagara*, East Indian Rosebay, Pinwheel flower, Nero's Crown, Moon Beam or

Carnation of India in different regions. It is a white coloured pinwheel shaped flower. *Tagara* plant is found extensively in the warm and humid regions of southern and eastern India. It bears pinwheel-shaped flowers. The plant canopy is of dark green colour and the leaves are glossy with glazed waxy coating. Crepe Jasmine plant grows up to a height of 3.2 meter. This plant is widely used as hedge plant. People prefer to plant it in home gardens due to minimum maintenance. In case the plant is grown in pot or container, pruning of roots and branches is to be done for better performance. *Tagara* plant is not eaten by animals, hence can be safely planted in boundaries as a decorative hedge. It has several medicinal uses.

Tagara plant is of two types *i.e.* single-flowered and double-flowered types. The flowers are white in colour. However, flowers of single-flowered variety do not have any fragrance whereas, double flowered variety has a mild scent.

Tagara is a tropical plant, hence grows well in warm and moist climate. The plant blooms abundantly during warmer months of the year. The plant growth is adversely affected due to low temperature during winter season when the plant goes into dormancy and sheds its leaves. Leaves of the plant become yellow and start to fall. It will start new growth as the soil becomes warmer during the spring and summer. The plants like sunlight and thrive very well in full sun or partial shade conditions. But the growth is retarded under full shade. *Tagara* can grow in any type of soil. However, it grows well in well-drained sandy loam soil. Improper drainage may result in root rot. The plant prefers slightly acidic soil, but it cannot tolerate alkaline condition.

This plant can be grown in any type of pot. However, earthen clay pot is suitable for plant growth. Earthen clay pot allows evaporation of water from the sides thereby reducing the risk of root rot. Potting mixture can be prepared by mixing 30% cow dung, 30% sand and 40% garden soil. Alternatively, a mixture of 30% cow dung, 20% sand and 50% garden soil may be taken to facilitate proper drainage. Now-a-days, ready-made potting mixture is also available in the market. Tagara can be easily propagated through stem cuttings. The cuttings start growing within 20-25 days. The best time to plant the cuttings is rainy season or spring season. The flowers grow in small clusters on the tip of the stem. Pruning of lower branch is necessary to maintain its shape and for accelerating plant growth. In case of potted plant, root pruning is necessary to facilitate better growth of canopy. Repotting is done by transferring the plants to a bigger size pot so as to accommodate more roots. Repotting should be done in the months of October-November. The plants can be prunned to accommodate in smaller size pots. Pruning of *Tagara* should be done preferably during the rainy season. However, pruning can also be done during September-October at the start of winter season.

All flowering plants need adequate quantity of nitrogen for growth of green leaves and phosphorus for flowering. A mixture of farm yard manure, dry leaf compost, neem oilcake and bone meal powder can be used to provide nutrition to the plant. Liquid fertilizer may be applied at an interval of 15 days to accelerate plant growth and flowering. Application of excess fertilizer in the potted plants may lead to scorching effect. The plant requires regular watering for better growth and development. Watering in excess quantity may adversely affect plant growth. *Tagara* plants are prone to attack by caterpillar, whitefly and scale insect. Neem oil or neem based pesticides can be sprayed to control the insects.

Yellow Oleander

Yellow oleander is widely grown as an ornamental plant for its attractive flowers. This plant is grown in the gardens, in pots or on the road side. The plant has thin green stems and leaves along with bright, shining orange or yellow flowers. Though it is a beautiful plant, the fruits and leaves are poisonous. It can reach up to a height of 2.8 meter. The plant performs well in bright sunny open space.

The leaves grow to a length of 10-12 cm. The upper side of the leaves are dark green, while the lower sides are pale green. This flower blooms during summer season and produces flowers in small clusters. The flowers are trumpet-shaped.

Yellow oleander is a hardy plant and can be grown in various types of soil. Proper drainage should be provided for potted plants. Regular potting soil in combination with compost may be used for better drainage. The plant requires adequate sunlight for desired growth and flowering. Yellow oleander is propagated through stem cuttings. Non-flowering stems are to be used as stem cuttings for effective survival of the plant. Cuttings of 10-12 cm length are used for propagation. Dipping the base of the cuttings in rooting hormone accelerates root formation and plant establishment. As this plant grows slowly, repotting is required in every 2-3 years.

The plants are to be provided with adequate nutrition at the beginning of flowering. A mixture of farm yard manure, dry leaf compost and neem oilcake can be used to fertilize the plant. If possible, balance application of nitrogen, phosphorous and potassium may be done to achieve good plant vigour and flowering. Yellow oleander requires adequate watering, especially during summer months. Thorough watering is essential during the blooming period. If the soil dries out, the blooms will wither away. The plants can be pruned before monsoon season to maintain good shape.

Yellow oleander is fairly resilient to pest or disease infestation. Scale insects and mealybugs are the most common pests affecting yellow oleander and feed on the leaves. Due to mealybug attack, there is development of cottony spots on the leaves. Scale insect leaves brown spots on the underside of the leaves. These insects can be managed by spraying insecticides or neem oil or neem based pesticide.

Anthurium

Anthurium is a colourful and long-lasting flower with attractive foliage. It is one of the major cut flowers in the world flower market. It is an important export-oriented crop with huge market potentiality. Anthurium requires humid condition and moderate temperature for successful flowering. The ideal daytime temperature for growing anthurium ranges from 25–28°C. Anthurium cannot grow well when the temperature is lower than 15°C or above 35°C. In tropical region, anthurium cannot be grown in the open space. Hence, shadenet with 75% shading is used for successful anthurium cultivation. However, the degree of shade varies depending on cultivar and local climate.

Anthurium can be suitably grown in well aerated soil, rich in organic matter with good water retention capacity and pH value of around 6.0. It is planted in beds or pots filled with well prepared growing mixture. Quality of growth media is directly correlated with production of high quality flowers of anthurium. The growing medium should properly anchor the roots for better plant growth. Presence of salt in the medium hampers plant growth. Usually, well-decomposed coconut husk or coco-peat, wood shavings, leaf mould, neem cake, cow dung, etc. are used as growing media for anthurium cultivation.

There are many anthurium varieties with different colours like red, orange, white, pink and bicolour. The red coloured varieties are the most favoured in the market. Some of the red coloured varieties are Aumana, Ayashii, Asahi, Cancan, Scarlette, Fla Red, Fla King, Mauritius Red, Tropical Red, etc. The orange coloured varieties are Nitta, Diamond Jubilee, Fla-range, Mauritius Orange, Horning Orange, etc. Some of the white coloured varieties include Manoa Mist, Ivory, Acropolis, Mauritius White, Lima, Deweese, Maunekea, etc. The pink coloured varieties are Aymara, Abe Pink, Blush, Surprise, Paradise Pink, Sarina, etc.

Anthurium is propagated by seed and vegetative means. The seeds germinate in 10 days and the seedlings are transplanted in the main field. Suckers are used for planting in vegetative method. The suckers with two to three good roots and at 4-5 leaf stage are separated from the main plant. Those are planted in pure sand for 3-4 weeks to develop appropriate root system. Application of

750 ppm benzyle adenine helps in increasing sucker production. Anthurium can also be propagated through auxiliary buds. An auxiliary bud along with a mature leaf and the root at the base is separated and planted in the moist sand. It is placed in a mist chamber till the bud develops to new plant. Now-a-days, tissue cultured plants are used for multiplication of anthurium.

The beds are prepared with a width of 1.20 meter and of convenient length. The slope of the bed is maintained for proper drainage. Plastic sheet is to be placed on the bed. The growing media are to be spread on the plastic sheet. The ideal growing medium consists of a mixture of coconut husk and cocopeat with a thickness of 20-30 cm. The growing media has to be disinfected by fungicides and pesticides to prevent infestation of disease and insect. The growing medium should be refilled after two to three years.

Anthurium is transplanted at a spacing of 30 cm x 25 cm in small holes at a depth of 5-6 cm. A handful of coir pith is put in the hole to avoid contact of plant root with the ground plastic cover. The beds are irrigated two days before planting. Usually planting is done during evening time and a sprinkled irrigation is provided immediately after planting. Anthurium can also be grown in earthen pots of 22-30 cm size. Re-potting should be done in every1-2 years. New pots should be of same size as that of the old pots. The plants need moist soil condition all throughout the growth period. Watering is done twice daily during summer and once daily during other seasons depending on environmental situation.

Application of plant nutrient is an important aspect of anthurium cultivation. Adequate quantity of organic manure in terms of farm yard manure, compost, cocopeat, vermi-compost, etc. can fulfil the nutritional requirement of the plant. It is required to apply two grams of NPK (17:17:17 or 19:19:19) per plant once a month for better flowering. Foliar application of 0.5 to 1.0 per cent water soluble fertilizer (NPK 17:17:17) will help better plant growth. At the time of flower initiation, application of nitrogen fertilizer should be reduced to enhance vase life of flowers. Besides, application of calcium is also important to obtain good quality flowers. Weeding is required for obtaining better plant growth. Pruning of leaves is important as too many leaves often lead to crooked stems and damaged flower buds. The old and diseased leaves are to be pruned regularly. Nearly 3-5 leaves are to be maintained on the plant to get good quality flowers.

Anthurium is affected by various diseases like bacterial wilt, anthracnose and leaf blight. Clean cultivation and use of disease free planting materials prevent disease infestation. Insects like white fly, thrips and mites affect the plant. White fly and thrips suck cell sap resulting in stunted growth of plant.

These insects can be controlled by spraying of imidachloprid or dimethoate. Mite infestation results in mottling and wilting of affected leaves. This can be controlled by application of dicofol.

Each plant produces about 8-12 flowers per year. The flowers are harvested with long stalks after unfolding of the spathe. Harvesting is preferably done during morning hours. The flowers are to be cut with a sharp knife, leaving 3 cm of stem on the plants to prevent rotting of the stem. After harvest, the basal portion of the flower stalk is to be kept in water to prevent drying. Flowers are washed with clean water to remove the dust and other particles to get a shining look.

Orchid

Orchid is a flowering plant with colourful blooms and belongs to the family Orchidaceae. Many orchid species have strong fragrance that attracts pollinators. Depending on growth habits, orchids are of two broad categories *i.e.* monopodial and sympodial. Monopodial orchids have a single erect stem with leaves arranged opposite to each other. The flowers appear from the base of the uppermost leaves. Orchids with this type of growth habit include the phalaenopsis and vandas. Sympodial is the common type of orchid, which grows horizontally with new shoots emerging from the rhizome. Leaves and flowers come out from the top of new shoots. Many sympodial orchids emerge from pseudo bulbs, which are swollen shoots. The pseudo bulbs store water and nutrients to help the plant survive during prolonged period of water stress condition. Sympodial orchids include cattleya, cymbidium, oncidium and dendrobium.

Orchids like *phalaenopsis* and *paphiopedilum* are suitable for humid tropical climate condition with day temperature of 23° to 30°C and relative humidity of 80 to 90 per cent. Warm-climate orchids including cymbidium and dendrobium perform well in temperature range of 15° to 21°C with continuous supply of moisture and adequate aeration. On the other hand, cattleya and oncidium types can be grown under dry climate. Masdevallia and epidendrum orchids grow in the cloud forests with average temperature range of 16° to 21° C and high humidity. Generally, orchids are light-hungry plants. It requires about 12 to 14 hours of light each day. The plants with few leaves or leathery leaves (cattleya and oncidium) need higher light intensity. But, the plants with soft and floppy leaves (*phalaenopsis* and *paphiopedilum*) are very sensitive to light. Hence, those should not be placed under bright sunshine. Most of the tropical orchids prefer a humidity range of 60 to 80 per cent. It is required to maintain humidity for growing orchids. Misting is beneficial for some orchids.

Terrestrial orchids grow well in the soil. But most of the tropical orchids are epiphytes, which prefer to grow in the air. The fleshy roots of orchids are covered with a layer of white cells. These cells act as a sponge to absorb water and protect the roots from heat and moisture stress. A growing medium suitable for orchid should provide good aeration and proper drainage along with support for good anchorage of the root. Peat moss, dried fern roots, sphagnum moss, rock wool, coconut fiber, lava rock, etc. can be used as medium for growing orchid. Some epiphytic orchids can also be tied with branches of tree.

Orchids can be propagated by division of the plant. Dividing a plant defers blooming for at least one year. The larger plants produce more flowers. Small size divisions require many years to mature. Orchids can be grown in relatively small pot as it requires less soil. Putting orchid in plastic pots makes it easy to re-pot the plant as the roots can be more easily detached. The potting mixture should facilitate good drainage. Orchids are placed over the pot filled with growing medium. The crown of the plant should be just below the top of the pot. Sometimes, it is helpful to use a piece of wire to secure the plant until its roots get established. Alternately, orchids can be tied with various media like coconut coir, wood pieces, etc. and hanged in the balcony, terrace or other places. Some orchids require re-potting every year, while others thrive well in the same pot for years together. Generally, orchids are not disturbed unnecessarily. Re-potting is done in case of degeneration of growing medium with reduced aeration and over growth of the roots or shoot.

Generally, orchids grown in home gardens are not provided with fertilizer. However, orchids grown on commercial scale are to be properly fertilized to sustain healthy growth. Water soluble fertilizer containing 17-17-17 or 19-19-19 grade of NPK can be used. Fertilizer should be applied during active growth stage of the plant. Orchids should not be fertilized immediately after re-potting. Misting the orchids with fish emulsion or seaweed extracts provides micronutrients to the plant.

Most of the orchids can tolerate water stress for a longer period, but it cannot survive excess moisture condition. Under excess water situation, the plants die because of inadequate air circulation. As a general rule, orchids are to be watered once a week. Water is applied only after the growing medium is dried out and care should be taken not to over water the plants. Orchids will not resume active growth after being re-potted. Watering is done sparingly during this transition period.

Normally, orchid plants suffer due to fungal diseases like foliar blight, leaf spot and flower blight. The plant growth is severely affected due to infestation of the disease. Application of fungicides like propiconazole, mancozeb or carbendazim can control these diseases. Sometimes, the plant is affected by caterpillars and aphids. Aphids suck the sap from the plant and the plant becomes weak. Spraying of imidachloprid or methyl demeton can control the inset attack.

Orchid is harvested when 60 to 70 per cent of the blooms open. In case of commercial cultivation, harvesting is done after complete opening of the flowers. The stem along with flower is to be cut with a knife at about 1 cm above the base of the plant. Depending upon types of species, vaselife of cut flowers varies from 18 to 56 days.

Post-harvest Handling of Cut Flowers

Cut flowers are highly perishable and susceptible to post-harvest losses. Specific post-harvest handling methods can enhance the shelf life of cut flowers.

Mode of harvest: Cut flowers are to be harvested by sharp cut with specialised tools. After harvest, those are to be placed in water or in a floral preservative solution. The angle of cut should be in slanting manner to absorb water through cut ends. Care is to be taken to avoid damage to the stems during harvest.

Conditioning of cut flowers: The flowers are to be kept loosely in a big container of water for free air movement around the petioles. It will restore turgidity of cut flowers after harvest. Flowers may be kept in the dark rooms to reduce water loss by transpiration. The cut end of flower is to be placed in floral preservative solutions. The floral preservative solution consists of citric acid (150- 400 ppm) along with silver nitrate (25 to 30 ppm) or silver thio-sulphate (25 to 30 ppm). Grading of flowers is done based on flower size, colour, quality and type of cut flower to get good price in the market. Wrapping is necessary for transportation of cut flowers. It is done by using polythene sheet, butter paper sheet, waxy sheets or newspaper. Packing of flowers helps in protecting the flowers from physical damage and minimizing water loss during transportation. Packing can be done in boxes, cartons, bins and other containers. Ideal package should restrict the process of respiration and transpiration.

4

Foliage Plants

Foliage plants are grown for attractive branches and leaves. These plants have several benefits such as improving air quality, promoting better mental & physical health and enhancing the aesthetic value of the immediate vicinity. Some of the foliage plants grown in indoor or in the garden are described in this chapter.

Dieffenbachia

Dieffenbachia is a foliage plant with large and showy leaves. This is native to the West Indies. It is a very popular plant grown in the home gardens due to easy maintenance. The beauty of the plant lies in the unique pattern of colour combination of the leaves. The leaves are yellow near the vein, which turns dark green towards the edge. This combination is random and even within the same leaf, different patterns of ribbons and spots of yellow & green are formed. It can be taken up under various light intensity and growing conditions. Dieffenbachia plants grow up to 2.4 meter height with wide leaves of around 50 cm long and 25 cm wide. This plant is poisonous and should be kept away from children. The sap of the plant is pungent and highly irritating for skin.

This plant doesn't require any particular soil type to thrive. It grows in almost all types of soil. The soil should be well drained and light textured. In case of potted plants, sandy soils can be suitably used after mixing with garden soil and farm yard manure in equal proportions. Dieffenbachia grows best when the temperature is between 18-23 °C. This plant attains optimal growth under bright and indirect sunlight. Diffused lighting under complete shade results in stunted growth. Exposure to direct sunlight burns the leaves. The plants placed in indoors should receive adequate light from all sides of the plant. It prevents the plant from growing towards lighted side.

This plant is propagated by stem cuttings or by using the suckers developed on the side of the plant. These planting materials are placed in moist sand until roots begin to develop and thereafter transferred to the pot or ground. This plant needs to have its root density checked annually. If the roots are dense, it has to be transferred to a bigger pot. Watering has to be stopped one week before repotting.

To ensure faster growth, balanced mixture of fertilizer may be applied to the plant in regular intervals based on visual appearance of the plant. Manures rich in nitrogen can be applied as food for dieffenbachia. Over dose of fertilizer may result in dropping of the leaves. Though dieffenbachia needs a moist soil all throughout its growing period, it cannot tolerate excess moisture. It can tolerate extended period of water stress, but the leaves will begin to shrivel.

Dieffenbachia suffers from various problems due to biotic and abiotic stress conditions. Browning of the bottom leaves is a common problem of the plant. These leaves are to be removed to keep the plant tidy and beautiful. Sometimes, the leaves appear bleached with a webby substance on the underside. This may be due to presence of spider mites. In this case, plants are to be sprayed with neem oil or neem based pesticide. If the plant is exposed to bright, indirect sunlight; there is production of blooms from the buds. The buds are to be removed immediately to maintain the plant growth.

Croton

Croton (*Codiaeum variegatum*) is a perennial evergreen shrub. This is a tropical plant grown for its colourful foliage. The leaves are thick, leathery and of varying colours, shapes and size. Croton is planted in outdoor gardens, hedges or indoor for the aesthetic value. Putting a croton plant adds to the colour of the background. As the plant grows older, the colour of the stem may change to darker shade. The generic word croton is often used to refer different species within the genus.

There are different types of croton with a range of shapes and colours consisting of red, orange, purple, pink, yellow, green and white. Some of the popularly planted croton varieties are banana croton, bush on fire croton, eleanor roosevelt croton, gold dust or gold sun croton, magnificent croton, mammy croton, Oakleaf croton, petra croton, sunny star croton, yellow iceton croton, zanzibar croton, etc. The sap of the plant can result in stain on clothes and cause skin irritation. Croton plants can grow up to a height of nearly ten feet. However, dwarf varieties grow to a lesser height. Usually, the croton plants are maintained at 1.2-1.6 meter height.

Croton plants can be taken in well drained sandy loam soil with good moisture holding capacity. The root growth is restricted due to excess water in the soil or growth medium. Being a tropical plant, croton prefers warm condition. If temperature drops below 13^0 C, the leaves may begin to turn brown. Croton performs well at a temperature range of 25-30 degree Celsius. This can either be grown in the indoors or in the outdoors.

Croton plants prefer full sunlight, but some species can tolerate partial shade. The colour of the foliage depends on the intensity of sunlight received by the plant. In order to attain a vibrant colour, the plants should be kept under good light. Croton plants kept indoors in containers should be placed in a way so as to receive adequate sunlight. Insufficient light results in fading of the colours of the foliage. To maintain the colour and glaze of the leaves, the plants should be regularly sprayed with water or wiped with a wet sponge to remove dust and other pollutants.

Croton plant is propagated through 8-10 cm stem cuttings. The cutting should have 3-5 leaves and can be planted in a small pot for root development. The planting materials are to be kept in a warm environment by maintaining adequate soil moisture. Under favourable conditions, the cuttings will develop root within a month and after which, those can be transplanted. With advancement in age, croton leaves develop colour. The plant should be supplied with farm yard manure and a balanced combination of nitrogen, phosphorous and potassium for better growth. In case of indoor plants, ideal pot mixture using equal proportion of garden soil, sand and organic manure is used for growing croton plant.

Being a tropical plant, croton grows well in a warm and humid environment. The root zone of the plant should remain moist, but not continuously wet during the plant growth period. In drier condition, misting helps to maintain healthy leaf growth. Croton plant requires frequent watering, but not to be over-watered. Excess water can cause root rot. Under water stress condition, new foliage starts wilting indicating the need of water. Croton plants need to be pruned at certain intervals to remove diseased and old growths. It helps to maintain desired shape of the plant. Dead leaves or branches should be removed from the base, but overgrown leaves or branches can be trimmed just above the node. More than one third of the stem should not be removed at a time.

Sometimes, the leaves of the plant wither due to various stress conditions attributed to movement of the plant from outdoors to indoors or *vice-versa*, imbalance nutrient supply, adaptation in a new environment, exposure to extreme temperature, insufficient light, improper watering or disease/pests infestation. The remedial measures are as per the diagnosis of the problem. Crotons thrive well in warm weather, but do not tolerate frost. During winter season, it is better to grow croton plants in containers placed indoors.

Generally, croton plants are not affected by insect or disease. At times, insects like aphids, mealybugs, scale insect, thrips, etc. infest the plant. These insects suck the cell sap from the plant causing stunted growth. Spraying of water can detach the insects from the plant. Various insecticides such as profenophos, imidachloprid, fenvalarate, etc. can be sprayed to control the insects. Besides, neem oil or neem based pesticides can be used to manage the pests.

Dracaena

Dracaena (*Dracaena marginata*) is an indoor foliage plant, which can be grown under various light intensities. The beauty and hardiness of the plant has made it a popular indoor plant. Dracaena has attractive aesthetic value and resilient to adverse climatic condition. The plant has elegant and unique look with a touch of exotic feeling in the living room, dining area or any other place. The leaves are long, slender and green with red colour margin. The potted plants can last up to a decade with proper care and repotting. There are various types of dracaena such as plain, colorama, bicolor, tricolor, tarzan, magenta, kiwi, etc.

Dracaena prefers well-drained sandy loam soil with pH ranging from 6.0 to 7.0. It is usually grown in pots filled with pot mixture consisting of equal proportion of soil, sand and compost. In order to enhance drainage, clay pebbles or small stones are to be placed at the bottom of the pot. It is necessary to go for repotting the plant in every 2-3 years. Gravels, small stones or pebbles may be placed under the beds to facilitate drainage.

It grows best with atmospheric temperature of 20 to 22°C. Dracaena can be placed near the window to receive abundant sunlight. Dracaena cannot be grown in dark places. This plant grows well under low light condition, but the growth is retarded. Exposure to direct sunlight can scorch the leaves.

Dracaena can be propagated by stem cuttings. Roots emerge from the stem cutting, but rooting hormone can also be used to accelerate rooting. The cuttings with root length of 2.5 cm are to be transplanted in a pot. These plants grow slowly, hence require minimum pruning. In case of vigorous growth of plant, the unwanted growth can be removed to give the plant a good shape. This plant requires fertilizer application for better growth. Household compost, vermi-compost and chemical fertilizer can be used to fulfil the nutritional requirement of the plant. Liquid fertilizer works well for dracaena plant. Used tea leaves and kitchen waste can also be used after fermentation to provide nutrition to dracaena plant.

Dracaena is drought resistant and doesn't require frequent watering. Hence, the plant is to be watered when the top soil is completely dry. Over watering may suffocate the plant roots. Application of water at 5 to 7 days interval is advised for better growth of the plant. Overwatering results in yellowing of leaves, while under watering causes drying of stems and dropping of leaves.

Dracaena is susceptible to spider mite and mealybug. Insects cause considerable damage to plant growth. Spraying of insecticides and neem oil can control the insects. Sometimes, dracaena is suffered due to attack of various fungal diseases. The diseases can be controlled by application of fungicides like carbendazim, mancozeb and propiconazol.

Aglaonema

Aglaonema plant is also known as Chinese evergreen or Philippine evergreen. It is a gorgeous green plant, mostly grown indoors for air purification. The leaf colour is variegated and having bright appearance, which enhances the aesthetic value of the room. This is an easy-to-grow houseplant being loved for its attractive colour and pattern. The colour of the leaf is green, pink, red, silver or yellow. The leaves are with stripes, speckles and gradients. The plant is evergreen and grows up to a height of about 50 cm. Aglaonema are generally slow growing plants. There are more than 100 varieties of aglaonema suitable for planting indoors and in gardens. The prominent cultivars are Red Peacock, George's Ruby, Harlequin, Anyanmanee, Black Lance, etc.

Aglaonema can grow well in soil with pH range of 5.6 to 6.5. Standard potting mixture with combination of garden soil, sand and compost in equal proportion can be used for indoor/potted plants. It can also be grown in container with clean water. Little bit of charcoal is added to the water for better growth. Fertilizer can be applied along with irrigation water once in a month. Over use of fertilizer may cause scorching of leaf.

Aglaonema is propagated through stem cutting. Healthy shoots of about 15 cm long with at least five leaves are removed from the mother plant and used for propagation. The cuttings are planted under indirect sunlight. The plant produces new roots in four to five weeks, after which the plant is ready for transplanting.

Aglaonema plants are flexible with regard to light requirement. The green varieties of aglaonema can tolerate comparatively low light condition, whereas bright light is suitable for the colourful species. The plant performs well during warmer part of the year.

Aglaonema grows well in moist soil condition and needs watering more frequently in warmer months. When the soil gets too dry, the leaves droop down and revive after watering. Over watering decelerates root activities resulting in yellowing of leaves. When the plant becomes overcrowded, it is to be transferred into a bigger container with fresh soil mixture. Care should be taken to avoid water stagnation in the pots. Adequate number of holes in the pot will facilitate drainage.

Repotting is to be done when the plant is overcrowded with roots and roots start coming out of the drainage hole. Repotting should be done during the spring or summer season at the active growing stage of the plant. For repotting, the diameter of the pot should be of 8-10 cm wider than the existing one. The plant can be removed along with soil around the root zone with the help of a garden trowel. This should be carefully placed in the new pot without disturbing the root.

Aglaonema is affected by various diseases. The plants may wilt due to fungal attack or over watering. Application of copper oxychloride or chlorothalonil can control the fungal diseases.

Areca palm

Areca palm (*Chrysalidocarpus lutescens*) is one of the most widely used plant for indoor and outdoor gardening. This is otherwise known as bamboo palm, butterfly palm, golden cane palm or yellow palm. The plant is big and bold with feathery and narrow fronds having nearly 100 leaflets. The plant grows about 15 to 25 cm per year. An adult plant can attain a height of 2 to 3 meter under indoor condition.

Areca palm can be trimmed without much effect on plant growth. The trimmed mature plants can be kept indoors up to ten years. However, adequate light is required for successful growth of the plant. Over exposure of the plant to direct sunlight may change the leaf colour to yellowish-green.

Areca palm grows well in fertile soil with slightly acidic reaction. The soil should be well drained. Waterlogging in the soil causes root rot in palm. Addition of sand and peat moss to the soil improves the soil texture, facilitates drainage and lowers the pH value. Well-drained, peat-based potting mixture is ideal for potted indoor plants.

Areca palm does well in average temperature range of 18-24^0 C. The plant growth is not affected, when the atmospheric temperature doesn't go below 10^0 C. It should be kept away from air conditioners and heat sources. Areca palm requires high humid condition for developing bright shinny leaves. Under excessive dry condition, the leaf tips turn brown.

Areca palm can be propagated by splitting the rooted suckers. Root division can be done at any time of the year, but the plant attains best of growth in hot and humid conditions. For propagation, a mature plant with multiple stems is selected and watered sufficiently to loosen up the roots. The palm is removed from the pot by patting the sides of the container. The roots are cleaned off the soil by rinsing with water. Four to five stems are detached from the parent plant with the help of a sharp knife. The divided stems are planted in a pot containing ideal potting mixture. The pot is placed in bright & indirect sunlight and watering is done.

Areca palm is a heavy feeder and requires adequate quantity of nutrient application through farm yard manure, compost, vermi-compost, liquid fertilizer and other sources. Excess application of chemical fertilizer to the potted plants can adversely affect the plant. Application of plant nutrient may be done after analysing the soil nutrient status. Nitrogen deficiency symptoms are seen on the older leaves. In such case, the leaves will have a light green colour with golden coloured petioles. Lack of potassium is noticeable especially in outdoor palm trees. Under potassium deficient situation, necrotic or translucent yellow to orange spots are seen on the leaves. These symptoms are noticed on the tips of the older leaves. Balance application of fertilizer is advocated to obtain better plant growth. Usually, potassium deficiency is not seen when grown in pots.

This plant requires moist soil condition, but it is very much sensitive to overwatering. It cannot tolerate waterlogging or water-saturated condition. Areca palms are sensitive to high level of fluoride in water. Watering can be done when the soil starts drying. This is important for outdoor plants during warm weather to keep the plant healthy.

Areca palm needs repotting in every two to three years. The plant growth is adversely affected due to overcrowded root growth in the limited space of the pots. The main reason for repotting is replacing the exhausted potting soil and to remove fertilizer salt deposits on the sides of the pot. The plant is to be taken up in a new pot with freshly prepared pot mixture. Deep planting may be avoided for better establishment and growth of the plant. As the roots are brittle, those should not be disturbed much while repotting. The soil should be properly pressed around the plant for tightly packing the root zone after filling the soil.

Areca palm is usually not much affected by insect or disease pathogen. Leaf tip burn is seen with yellowing of the tips and leaves. This may be due to water stress condition or infection of bacteria. Bacterial infection can be controlled by application of plantomycin. Sometimes, aphids attack the leaves and suck the cell sap from the plant. Those can be managed by application of neem oil or neem based pesticides. Cleaning the area around the plant reduces the risk of disease and insect attack.

Fern

Ferns are considered as one of the oldest plants, dating back to prehistoric ages. These plants can thrive well both in indoor and outdoor condition. Ferns are very good oxygen supplier, air purifier and natural humidifier. Ferns are popular among the gardeners due to lush green foliage, easy maintenance and air purifying properties. Due to its ancient origin, ferns have thousands of varieties. Ferns are essentially tropical plants and grow well under warm and humid condition. The plant requires little care for growth and development. Some common but stunning fern varieties are Grey Ghost Lady Fern, Hart's Tongue Fern, Staghorn Fern, Hen and Chicken Fern, Kangaroo Fern, Bird's Nest Fern, Painted Lady Fern, Asparagus Fern, Boston fern, Giant Wood Fern, etc.

Ferns need little management and care. Ferns do not require specific growing medium. Ferns grow well on the layer of fallen leaves, tree bark and decaying matter. These can be taken up in loose and well drained growing media. A suitable medium can be prepared by mixing soil, coco peat, perlite/pumice and coarse gravel. Since ferns grow under canopies of tree, it is best to place the plants in an area with indirect sunlight. Care should be taken to keep away the plant from direct sunlight. The soil should always be kept moist for suitable plant growth. Ferns cannot be grown under dry condition as it prefers humid condition. Excessive drying results in browning of frond tips. Ferns can be kept as hanging plants in bathrooms with good light due to their preference for humid environment.

Application of water soluble fertilizers through foliar spray is preferred for a healthy canopy growth in ferns. Over dose of fertilizer may cause injury to the plant. Plant nutrients can be provided through application of organic matter such as farm yard manure, compost, vermi-compost, vermi-wash, coco peat, etc. The dead fronds should be removed in regular interval to give the plant a vibrant look.

Exposure to direct sunlight for a longer period, over watering or water stress condition results in browning of leaves. Overwatering or soggy soil causes crispy and brown leaves in the centre of the plant. Use of well-draining potting mixture can avoid waterlogging and promote plant growth. Inadequate humidity in the atmosphere results in browning of leaf tips. Water stress condition often causes dropping of leaves.

Caladium

Caladium has become one of the most popular foliage plant under shade area of the gardens. The lush multi-coloured leaves make this plant attractive and eye catching. The leaves bear various patterns of striking colour combinations of green, white, pink and red. The plant thrives well in hot and humid conditions. Caladium grows up to a height of 45 to 60 centimetre. Based on the type of leaves, caladium is of two main types such as fancy-leaf and strap-leaf. The fancy-leaf caladiums usually have large heart-shaped leaves growing to more than a foot long. Whereas, strap-leaf varieties have narrow leaves with compact foliage. There are various types of caladium grown by the gardeners. Some of those are Aaron Caladium, Carolyn Whorton, Candidum, Freida Hemple, June Bride, Miss Muffett, Postman Joyner, Rosebud, White Queen, Apple Blossom, etc.

Caladium can be planted as groundcover, border plant or in container to impart the colour effect in the garden. The large colourful leaves attract attention in the garden. These plants can be placed in corner of a room or in the backdrop of other shade loving plants.

Caladium is propagated through tubers. Caladiums are tuberous plants and are propagated by small tubers separated from the root system. These can be cut into pieces with at least one eye or knob on each piece or can be planted as whole tuber. The tubers are planted directly into damp soil. The tubers or the cut pieces are dried for a few days before planting. Large tubers produce more leaves and mature earlier as compared with small ones. The plant grows well in fertile and well-drained soil supplied with adequate organic manure. It requires partial or complete shade to have vibrant leaf colour. The plants having larger leaves are to be placed in such a point in the garden so as to avoid damage due to strong wind.

Being tropical plant, caladiums flourish in warm climate with daytime temperature ranging from 21 to 30° C and night temperatures above 16° C. The best time for planting caladium is from March to June. Under cooler climatic condition, caladiums perform better in containers as the soil warms up faster. The tubers are planted with plant to plant distance of 20 to 30 cm at a depth of 4 to 5 cm. The pointed end of the tuber is to be placed upward. The potted caladium needs to be acclimatized to the outdoor conditions before putting those in the ground. Application of plant nutrients helps in better growth and development of leaf colour in caladium. Balance application of major nutrients along with adequate bulky organic manure result in vigorous plant growth. Excessive use of chemical fertilizer can be detrimental for the plant. The plant requires moist soil condition throughout the growing season to remain vibrant. Water stressed soil condition results in yellowing of leaves and causes leaf drop.

Sometimes, caladium plants are affected by caterpillars and aphids. The symptoms of insect attack are appearance of ragged margins on foliage. The infestation can be controlled by use of *Bacillus thuringiensis*. Caterpillars can be handpicked to protect the plant. The diseases of caladium plant are usually seen on the tubers. The diseases are caused by the fungi *Rhizoctonia* and *Pythium* species. These can be controlled by spraying of neem oil or neem based pesticides.

Coleus

Coleus is one of the most preferred foliage plant in the garden for its colourful leaves. Special features like tolerance to direct sunlight, prolific branching and compact growth habit have made this plant popular among the gardeners. The leaves have variety of patterns with solid colours, splashes, blotches, streaks, flecks and veins of attractive colours. The colour intensity varies according to the exposure of the plant to sunlight. The varieties with deep coloured leaves can tolerate more sunlight, while the leaves of light colour varieties suffer from scorching when exposed to direct sunlight for longer period. Plants require semi-shaded condition or indirect sunlight to maintain consistent colour of foliage. Plant grown under little light becomes less vigorous and the improper expression of colour. Coleus plants have various leaf texture such as large, small, twisted, elongated, scalloped, lobed, finger-like, etc. These plants are of three types *viz*. upright, rounded and trailing.

The leaf colour of coleus is better achieved with exposure to morning sun and some degree of afternoon shade. Many varieties perform well both in shade and semi-shade condition. Some of the varieties do well under direct sunlight as long as those are not allowed to dry out. The plant prefers a well-drained fertile soil. The damp soils may cause leaf drop and encourage disease infestation in the plant. Coleus prefers warm temperature above 16 degree Celsius. Some of the commonly grown coleus varieties are Rustic Orange, Fishnet Stocking, Wizard Mix, Henna, Black Dragon, Picture Perfect, Salmon Pink, Big Red Judy, Wizard Jade, Wizard Velvet Red, Premium Sun, Crimson Gold, etc.

Coleus grows well both in raised beds and pots. The potting mixture should be nutrient rich and well drained. The pots are to be manured regularly and placed in such a way so as to receive adequate sunlight. The soft twigs and leaves of coleus are prone to damage when exposed to high wind. Coleus can also be grown in hanging baskets. Some of the trailing varieties can grow over the edges, while compact and bushy types can be planted in the middle of the basket. Coleus can be propagated from seed and stem cutting. The seeds are sown in pots and kept under bright sunlight for better germination. However, this can also be propagated with ease by use of stem cuttings.

Addition of organic manure will help satisfactory plant growth. The potted plants are to be fertilized regularly with water-soluble fertilizer. If over-fertilized, coleus may not express proper colour. So, adequate care should be taken during fertilizer application. Coleus is a thirsty plant and needs regular watering. Overwatering leads to discolouration of leaves and stunted growth of the plant.

The plant needs pinching to maintain an attractive form of the foliage. Regular pinching prevents flower formation. Pinching is done just above a set of leaves or branching junction to obtain good appearance of the plant. Pinching helps in diverting more plant nutrients and energy for vigorous growth of stem and foliage. Allowing the plant to flower at early stage may result in loss of vigour due to diversion of photosynthates towards flower and seed development.

Coleus plants are generally tolerant to pests and disease infestation. However, occasionally these plants suffer from disease and pest problems. Waterlogged or flooded soil can result in root rot and stem rot diseases. The leaves of the affected plants turn brown and eventually the plants die. The plant is also affected by the fungal disease downy mildew. Tiny spots appear on the foliage and there is leaf distortion and dropping of leaves. Downy mildew is more common in greenhouses under high humidity. These diseases can be controlled by spraying of fungicides such as copper oxychloride or carbendazim. Coleus plants are sometimes attacked by mealy bugs, whiteflies, aphids, etc. Pest infestation causes spots and holes in leaves and stunted growth of plant. This can be controlled by spraying of insecticides like profenophos, imidachloprid, etc. The disease and insects can also be managed by application of neem oil or neem based pesticides.

Snake Plant

Snake plant (*Dracaena trifasciata*) is one of the hardiest species of household plants. It is an evergreen perennial plant. The leaves are stiff with sword-like appearance. Generally, the leaves have green-bands with yellow border. It is easy to maintain snake plant in the garden. Snake plant is not sensitive to light intensity and can thrive in bright light or dark condition. However, exposure to bright light enhances the growth process as compared to low light condition. The plant attains a height up to five meter. Snake plant grows well in containers or pots. The average lifespan of a snake plant is five to ten years. However, it can live up to 25 years with proper care and maintenance. There are various types of snake plants.

- *Dracaena trifasciata* is otherwise known as Hahnii or bird's nest snake plant. This is a dwarf type plant, which grows to a height of 15 cm. The leaf arrangement looks like a bird's nest.
- *Dracaena angolensi* is a cylindrical snake plant with round and stiff long leaves. The leaves curve outward from a central crown.
- *Dracaena trifasciata* or *Laurentii* is a variegated snake plant. The leaves have creamy yellow margins.
- *Dracaena trifasciata* has twisted leaves. There are horizontal stripes on the leaf with yellow coloured edges. The plant grows up to a height of 35 cm.
- *Dracaena trifasciata* is a type of snake plant growing up to a height of 1.2 meter and has narrow leaves with white vertical stripes.
- *Dracaena pearsonii* is also known as rhino grass. It grows up to 30 cm in height with succulent red-tinted leaves.

Snake plants prefer well drained light or sandy soil. Loose and well-drained potting mix should be used in pots. The potting media should contain less quantity of peat as it may lead to ill drainage. Snake plant prefers indirect and steady sunlight. It can adapt to full sunlight and can survive under partial shade. Snake plants prefer warm condition and suffer when exposed to temperature below 10 degree Celsius. Ideal temperature ranges for growing snake plant is between 21 and 32 degree Celsius. The plant cannot tolerate frost.

Snake plant is propagated by dividing the plant when the height is at least 10 cm. Alternatively, the new shoots emerging from base of the plant can be used for propagation. It can be propagated through root division and leaf cutting. In case of root division, the plant is separated with roots using a sharp knife, keeping the roots intact for each section. These cuttings can be used to grow new plants. The new offshoots developed from the base of the old plant can also be used to propagate snake plants. Leaf cuttings of snake plant are also used for propagation. In this method, a long and healthy leaf is sliced off the plant using a sharp knife or pruning shear. The leaf cutting is dipped in a container filled with clean water. The pot is placed under partial shade for root development. Once the roots become 2-3 cm in length, the leaf cutting is ready for transplanting in to a pot containing well prepared pot mix. Alternatively, the leaf cuttings are allowed to harden for 24 hours and planted in pots.

The plant is to be provided with adequate plant nutrients in the form of farm yard manure, vermi-compost and chemical fertilizer for better growth and development. The growth media should contain about one-third of organic manure for better performance. Water soluble fertilizer can also be sprayed on the plant at certain time interval to augment plant growth. Overwatering is harmful for the plant and causes root rot. Watering is done based on requirement of the soil.

Damaged or over matured leaves are to be removed by use of pruning shears, scissors or sharp knife to facilitate new growth. Rainy season is the best time to prune the plant. The tallest leaves of the plant are removed to maintain proper height of the plant. Being a slow grower, snake plant doesn't require frequent repotting. However, repotting can be done to replace the exhausted pot mixture or transferring the plant to bigger pots.

Snake plant is attacked by various insects like thrips, aphids, mealybugs and whiteflies. Leaf curling is caused due to attack of thrips. The aphids suck the cell sap resulting in stunted growth of the plant. Insect attack can be avoided by keeping the plant healthy. Addition of neem oilcake to the pot mix or spraying of neem based pesticide or neem kernel extract can keep the insects away from the plant.

Snake plant leaves grow upward and stand erect. But exess water, insufficient light or poor potting material can make the leaves droop. Soggy soil also adversely affects the foliage growth. The plants remain healthy under bright sunshine, judicious watering and proper nutrition.

Cactus

Cacti and succulents are attractive plants for landscape gardening. Cacti are a group of xerophytic plants with various shape and size. Being grown in the water deficient situation, the plant has unusual appearance with typical morphological, anatomical and physiological features so as to adopt under such condition. These are perennial herbs or shrubs characterized by fleshy stems and clear cut joints. The leaves are generally absent or rudimentary or reduced to mere minute scales. Cacti require minimum maintenance and care. These plants can be grown on well prepared bed, pots or containers. Cultivation of cactus is called cactus farming. Now-a-days, many agri-entrepreneurs are taking up cactus farming in commercial scale. There are many uses of the edible cactus plant, which make commercial cactus farming a profitable business. Cacti vary in size, shape, colour and growth habits. The appearance can be in the shape of upright columns, spreading clumps or spiny balls. Cacti are available in numerous varieties. Some of those produce attractive flowers. Various types of cacti are discussed here.

Beavertail cactus: Beavertail is a prickly pear cactus with blue-green coloured pads. The plant attains a height of about 50 cm and spreads up to 180 cm. The flowers are dark cherry pink in colour and are ideal for water scarce condition. When planted on ground, the plants hardly need fertilizer. However, potted plants may be supplied with water soluble fertilizer once in a year.

Blue flame cactus: Blue flame cactus is known as bilberry cactus. It can grow up to a height of five meter and spreads over an area of 2-3 meter. Blue flame cactus forms dense vegetation. This cactus looks beautiful, when placed in pots filled with gravel mix and rocks. It can be the point of attraction in a garden developed under water stress condition.

Golden ball cactus: Golden ball cactus is also known as lemon ball cactus or yellow tower. It grows about 1.2 meter height and forms in clusters. It is an excellent choice to start cactus gardening. Golden ball cactus requires partial shade during summer days.

Golden barrel cactus**:** Golden barrel cactus can grow up to 1.6 meter height. This cactus is popular in water deficient areas. Planting may be done in lines to obtain attractive visual effect. The plant requires light shade at early growth phase, but the mature plants grow well under full sunlight.

Strawberry hedgehog cactus: Strawberry hedgehog cactus is also known as strawberry cactus or purple torch. These cacti are small in size with 70 centimetre height and grow with free-branching clusters. This cactus looks attractive in rock and water deficient garden. It requires light shade for better growth.

Organ pipe cactus*:* The stems of this cactus resemble with sausage. The tall and columnar plants branch into several ribbed stems. Organ pipe cactus makes an attractive appearance in the garden. It can also be used as hedge plant.

Peanut cactus*:* Peanut cactus is a cylindrical cactus, which grows up to a height of 30 centimetres. Its vase-shaped blooms appear during early summer season. Peanut cactus prefers shallow containers.

Prickly pear cactus*:* The size of flower of Prickly pear cactus is about 10 cm. The cactus is easily identifiable by its red or yellow decorative fruits. It can grow up to five meter tall and spreads up to 2.4 meter wide. Prickly pears are suitably grown in water deficient gardens. Prickly pear cactus rarely requires fertilizer supplementation under outdoor condition.

Rat tail cactus : Rat tail cactus is suitable for hanging purpose. The potted plants can grow up to a height of two meter. This cactus produces beautiful tubular flowers.

Silver torch cactus: Silver torch cactus grows in the form of slender columns and can attain a height of 3.2 meter. The flowers are tubular and protrude horizontally. The plant performs well under bright sunlight and in well-drained soil.

Star cactus: Star cactus can grow up to a height of one meter. It is identifiable by five to eight ribs twisting in to spirals. It has bright yellow-brown spines and shiny yellow flowers. It can be grown both on the field and in wide mouth container.

Cacti can be grown under wide range of temperature. Extreme temperature adversely affects the growth of cacti. Temperature above 35^{0}C may have scorching effect on the plants. Hence, the plants should be kept under shade or indoor condition to have better performance. Cactus requires porous sandy soil, which has excellent drainage and it prefers calcareous soil. Finely powdered lime mortar from old buildings, hard lime stones, marble, crushed shells of oysters or other shell fish, etc. can be added to the soil before planting. Leaf mould, compost, well-rotted cow dung can be added to the growing medium for providing nutrition to the plant. Use of undecomposed organic matter may be avoided as it retards plant growth. Small quantities of bone meal, superphosphate or basic slag may be added to the compost. Addition of charcoal powder also helps to keep the soil porous. Cactus can be suitably planted during the active growth stage. The soil should not be too wet. In case roots are damaged during planting, water supply is to be stopped for few days for recover of the roots. Subsequently, watering may be done to facilitate new growth.

Cacti can be propagated by cutting of offset. Removal of the top portion facilitates development of side shoots. These side branches can be used for propagation. The branches are cut with a sharp knife and kept for few days in well ventilated and shade area for hardening. The cutting is then planted in sand and leaf mould or in the compost for root development. During the period of root formation, watering is done cautiously. Formation of roots in the cuttings is indicated by new growth. Some species are best propagated by separating the suckers arising from the lower part of the stem.

Cacti can also be propagated by grafting of plants. A clean horizontal cut is made on both stock and scion for grafting. The scion is fixed on the rootstock by giving moderate pressure. A piece of cloth is placed on the top of the scion and the string is tied to avoid injury. The cut end of the stock and scion may be inclined at an angle of 45 degree in case of tongue graft. Grafting of cacti can also be done by inserting the scion in the stock. Grafting should be done during dry season when the humidity is very low as high humidity causes rotting.

An ideal pot mix can be prepared by mixing garden soil, sand, old mortar or limestone gravel and leaf mould. Earthen pots are suitable for planting cacti as ceramic or plastic pots hinder aeration of the growth media. It is advisable to use pots large enough to accommodate the plant. Pots retaining too much moisture may be harmful for the plant.

The plants are to be kept dry for repotting so that it comes out easily without injuring the roots. The plants should be replanted just as deep as in the previous pot. Newly potted plants are to be watered once or twice a day and placed in shade for about one week. Cacti are xerophytic plants and adapted to drought condition. The physiological condition requires less water. During summer months, watering is done more frequently. However, it has to be ensured that the soil is drained properly. Minimum watering is required during winter months as the roots remain inactive. Spraying of water in the morning is good for cacti. There should not be accumulation of water in the depression of the plant as it may lead to rotting.

Cacti suffer from various fungal diseases like anthracnose, charcoal rot, dry rot, etc. There is appearance of discolouration of plant and rotting due to disease attack. The plants may die due to fungal disease. These diseases can be controlled by spraying of fungicides like propiconazole, copper oxychloride or carbendazim. Insects like mealy bug and thrip pose threat for cactus. The insects suck the cell sap and hinder the plant growth. These insects can be controlled by application of imidachloprid or profenophos. Application of neem oil or neem based pesticide can control insect and disease infestation.

5

Fruit Plants

Fruit trees can be planted in the backyard or any open space near the house. In absence of backyard area, fruit trees can also be grown on the rooftop by planting in polythene bag, plastic container, cemented pot, etc. Growing fruits like papaya, banana, lemon, mango, pineapple, coconut, pomegranate, etc. in the premises of house can fulfil the family requirement of fruits. It will help to get fresh fruits from own garden, which are free from any chemical or pesticide. There is a different satisfaction in plucking of fruits grown by own effect.

Mango

Mango (*Mangifera indica*) is one of the prominent fruit crops of India. It has delicious taste and excellent flavour. Mango fruit is utilized at all the stages of its development. Raw fruits are used for preparation of pickle, chutney and juice. The ripe fruits are consumed directly and are used for preparation of several products like squash, syrup, jam and jelly. Ripe mango contains about 0.6 per cent protein, 0.1 per cent fat, 11.8 per cent carbohydrate, 0.3 per cent minerals and 1.1 per cent fibre. Mango can be grown both under tropical and sub-tropical climate. Good rainfall and dry summer are favourable for mango cultivation. This plant can be grown in a wide range of soil starting from alluvial to laterite soil. The ideal soil pH should be in between 6.5 and 7.5 for better plant growth and yield. In the rooftop, it can be planted in container, polythene bag or gunny bag filled with mixture of soil, compost and sand.

There are various varieties of mango available for cultivation. Some of the popular varieties are Amrapali, Mallika, Neelam, Alphonso, Rumani, Banganapalli, Kalepad, Jahangir, Mulgoa, Himayuddin, Bangalora, Arka Anmol, Arka Aruna, Arka Neelkiran, Arka Puneeth, Totapuri, Bombay green, Himsagar, Sindhu, etc.

Mango is propagated by stone or grafted seedlings. The grafted plants can be procured from a reliable source in the local area. The graft should be healthy and disease free. Planting is normally done during the month of July. However, planting can be done in other months depending on local climatic condition and availability of irrigation facility. The tree is planted in pit with size of one meter long, one meter wide and one meter deep. The pits are prepared at a distance of 7-10 meter depending on variety. In case of high density planting, the seedlings are planted at a spacing of 5 meter X 5 meter, thereby accommodating about 400 plants/ha. Each pit should be filled up with topsoil mixed with 10 kg of FYM and 100 g neem oilcake. Grafts are planted in the centre of pit with ball of soil intact around its root. While planting, care should be taken to keep the graft union point 15 cm above the ground level.

Fertilizers are applied 45 to 90 cm away from the trunk and incorporated in the soil. The quantity of fartilizer depends on soil test results and method of planting. However, general recommendation can be followed in absence of soil test values. During first year, application of 20 kg FYM, 200 g N, 200 g P_2O_5 and 300 g K_2O is to be done per plant per year. The dose has to be enhanced every year till 6th year of the plant. After 6th year, every plant is to be provided with 50 kg FYM, 1.0 kg N, 1.0 kg P_2O_5 and 1.5 kg K_2O per annum. Manures and fertilizers are to be applied in two equal splits in a year. The first dose is to be applied during June-July and second dose during October. If planted in pots, container or polybags, chemical fertilizer can be applied in more splits to have less concentration of chemicals in the container. Application of micro-nutrients like zinc, copper and boron prevents fruit dropping and improves flowering & fruiting. A mixture of 50 gm zinc sulphate, 50 gm copper sulphate and 20 gm borax is recommended per tree per annum to obtain good yield. In case of plants grown in container or polythene bag, care should be taken to avoid scorching effect on root or leaf due to fertilizer application. Under such situation, preference is to be given for application of organic manure like farm yard manure, compost, vermi-compost, oil cake, etc. Water soluble fertilizer can also be applied to avoid the risk of scorching.

The plants need to be irrigated at regular intervals till establishment. Generally, irrigation is given at weekly interval. The trees are to be pruned during rainy season, once in three years. The trees are not allowed for fruiting up to three years by removing the inflorescences. In case of container planting, the plant is to be trimmed frequently to reduce the foliage. Growth regulators induce flowering and fruiting in mango. Spray application of NAA @ 20 ppm at flowering increases fruit retention.

Mango trees are affected by various insects and diseases. Mango hopper is a common insect affecting the tree. The symptoms of mango hopper are shedding of flower buds, presence of honey dew secretion on lower leaves and development of sooty mould. This can be controlled by two sprayings of acephate 75 SP@ 1g/lit or phosalone 35 EC @ 1.5 ml/litre water. Neem oil @ 5 ml/litre of water can be sprayed for control of the hopper. Sometimes, mites cause damage to the plant. The mites appear as white coating on the upper leaf surface, which can be easily rubbed off by hand. Subsequently, the leaves turn yellow and drop prematurely. Spraying of wettable sulphur @ 2 g/lit can control mite infestation.

Nut weevil insect causes reddish-brown water-soaked spots on the fruits. The affected fruits have hole in the stones and the inner core of the fruit turns into black coloured rotten mass. Spraying of fenthion 100 EC @ 1ml/lit twice can control nut weevil. Mealy bug sucks the cell sap adversely and affects the plant growth. Spraying of Profenophos @ 2 ml/lit can control the pest. Fruit fly is another harmful insect of the tree. The maggots bore into semi-ripen fruits resulting in fruit dropping. This can be controlled by spraying of Fenthion or malathion@ 2 ml/litre of water. A solution of jaggery and malathion may be kept under the tree to kill the insects.

Mango is infested by powdery mildew and anthracnose disease. The typical symptom of powdery mildew disease is appearance of white powdery fungal growth on leaves, branches and young fruits. The affected fruits drop prematurely. Application of sulphur dust or spraying of wettable sulphur can control powdery mildew disease. Symptoms of anthracnose disease occur on leaves, twigs, flowers and fruits. Small, angular, brown to black spots are seen on the leaves, which enlarge to form dead areas. Pre-harvest spraying of Thiophanate methyl @ 1g/lit or Chlorothalonil @ 2 g/litre can control the disease. Sometimes, sooty mould disease affects mango plant. The fungi produce mycelia, which grow on sugary secretions of the plant hoppers, jassids, aphids and scale insects. Control of insects by spraying profenophos @ 2 ml/litre or other insecticides can minimize the risk of sooty mould disease.

Mango is harvested when the fruit is fully mature. On an average, one tree can produce 70-80 kg fruits per year. But, the plants grown in container, pot or polybags may yield less fruits.

Banana

Banana is a popular fruit consumed at raw stage as vegetable and at ripen stage at fruit. Banana is basically a tropical crop and grows well in temperature ranging from 15°C to 35°C. This crop can be cultivated from humid tropical to dry mild subtropical climate. High velocity of wind damages the crop. Deep and loamy soil with pH ranging between 6.5 -7.5 is suitable for banana cultivation. Soil should have good drainage, rich in organic matter, adequate fertility and moisture content for growing banana. Saline and calcareous soils are not suitable for banana plant. Banana is grown for culinary and desert purpose. The desert types are Robusta, Dwarf Cavendish, Grand Naine, Rasthali, Poovan,

Nendran, Red Banana, Karpooravalli, Co-1, Matti, Sannachenkadali, Udayam and Neypoovan. The culinary purpose varieties are Monthan, Ash Monthan, Bantala and Chakkia. Nendran is a dual purpose variety used for dessert and culinary.

Banana is commonly propagated by sword suckers and water suckers with narrow leaves. Sword suckers should be free from diseases and nematodes. The corm of the sucker is to be dipped in chloropyriphos and carbendazim to prevent disease and insect attack. Tissue cultured banana plants with 5-6 leaves can also be used for propagation. At the time of planting, application of 25 g *Pseudomonas fluorescence* per plant is beneficial. Application of *Azospirillum* and Phosphobacterium @ 20 gm each per plant may be done at the time of planting. Planting of banana is done by pit method or furrow method. Planting is generally done during June-July. Planting can also be done at any time of the year depending upon local climatic condition and availability of irrigation facility. The planting distance is 3 m X 3 m for tall varieties and 2 m X 2 m for dwarf ones. Planting is done in the pits of size 50 cm X 50 cm X 50 cm. Planting can also be done in container, polythene bag or gunny bag placed on the rooftop or terrace. In such case, adequate quantity of mixture containing soil, sand and farm yard manure is to be filled in the container or the polybags to support growth of the plant.

Banana is a quick growing plant and requires adequate nutrition. Banana responds well to manures and fertilizers. Basal application of 10 kg FYM or compost and 100 g neem oilcake is done per plant before planting. Application of 100 g N, 30 g P_2O_5 and 300 g K_2O per plant is to be done in three splits during second, third and fifth months after planting. The fertilizer should not come in direct contact of the plant at the time of application. Spraying of micronutrients like $ZnSO_4$ (0.5%), $FeSO_4$ (0.2%), $CuSO_4$ (0.2%) and H_3BO_3 (0.1%) at 3, 5 and 7 months after planting can increase the yield and quality. Care should be taken during application of fertilizer for the plants grown in container or polythene to avoid scorching effect. In such case fertilizer can be applied in small quantity to avoid adverse effect on the plant. Irrigation may be done immediately after planting. Subsequent irrigations are to be given once in a week or depending on soil moisture status. The field should be made weed free for better plant growth. Earthing up helps in better drainage and facilitates irrigation.

Desuckering should be done at monthly interval to keep one plant per pit. The dry and diseased leaves are to be removed to prevent spread of disease. Male flowers may be removed a week after opening of the last hand. The plants at flowering stage may be given support by bamboo or wooden pole.

The peduncle may be covered with flag leaf to prevent stalk end rot. Covering the bunch with banana leaves or polythene can protect the fruits from sunscald and will improve appearance of the fruits. The quality of the fruits can be increased by use of plant growth regulators. Application of 2,4-D @ 25 mg/litre will improve the quality of bunches.

Banana is affected by various insects like corm weevil, stem weevil, aphid, etc. Corm weevil makes hole in the corm, which leads to rotting of corm and wilting of plants. Cleaning the field, de-suckering, removing undesired leaves can reduce the attack of corm weevil. It can be controlled by application of neem oil, neem based pesticides and chemical insecticides. Stem weevil makes hole in the stem, which results in poor plant growth. Removal of dried leaves and keeping the plantation clean prevents spread of the insect. Chemically, it can be controlled by application of profenophos @ 2 ml per litre of water. Banana is also affected by aphids, which act as a vector for bunchy top virus disease. Spraying of Methyl demeton @ 2 ml/lit or Dimethoate 30 EC @ 2 ml/lit can control this insect.

Sigatoka leaf spot is a harmful disease of banana. It can be managed by removal of affected leaves. Spraying of fungicides like carbendazim or Mancozeb or copper oxychloride or propiconazole can control this disease. Bunchy top is a viral disease spread by aphid. The plant growth is inhibited and yield is reduced due to infestation of this disease. Use of disease free suckers and control of aphid can prevent spread of this disease. Panama wilt is another serious disease associated with banana. The affected plants are to be uprooted and destroyed. Removal of the affected plants and application of lime @ 1-2 kg per pit can control the disease. Spraying of Carbendazim @ 2g/litre of water can also control this disease.

Depending on variety and local climatic condition, bunches are ready for harvest after 12 to 15 months of planting. The fruits attain maturity from 100 to 150 days after flowering. A banana plant can produce about 9 to 12 hands with 14 to 20 fingers per hand. In this way, a single banana plant can produce up to 240 bananas. A single plant can yield about 20 kg fruits per harvest. The plant can be maintained up to three years in the same pit.

Coconut

Coconut is a very beneficial fruit tree as each part of the plant is useful. It provides food, drink, fuel and organic manure. Coconut fruit is commonly used for its water, milk and oil. Coconuts are especially high in manganese, copper and iron. It can be grown under varying climatic and soil conditions. It is a tropical plant, which can be grown in the temperature range of 22 to 32°

C. The plant grows up to 600 m above mean sea level. Plantation of one or two trees in the backyard is very much helpful to get coconut to fulfil family requirement. As coconut requires less area, it can be taken up by urban families with limited area available. This can be planted near the boundary or corner of the house premises.

Coconut varieties are classified into tall and dwarf groups. The tall cultivars include West Coast Tall, East Coast Tall, Andaman Ordinary, Philippines, Java, Cochin-China, Kappadam, etc. Some of the dwarf varieties are Chowghat Green Dwarf, Chowghat Orange Dwarf, Malaysian Dwarf, Chavakadan Dwarf, Sunnaki Dwarf, etc. The promising hybrids are Tall X Dwarf (TxD) and Dwarf X Tall (DxT). Coconut is propagated through seedlings raised from nuts. The seed nuts are to be collected from healthy and good yielding plants. Matured nuts of about 12 months old should be selected. Nuts of too big or too small in size or of irregular shape should be discarded. Generally, 9 to 12 month old seedlings having 6-8 leaves are used for planting. Good quality seedlings can also be obtained from the nurseries.

Deep soil with good water holding capacity is preferred for coconut cultivation. Shallow soils with underlying hard rock or low lying areas are not suitable. Alluvial, red, sandy loam soil with pH ranging from 5.5 to 8.0 is suitable for coconut planting. Pits of 1 m length, 1 m width and 1 m depth may be dug for planting. The spacing should be 7.5 m X 7.5 m for tall varieties and 5.0 m X 5.0 m for dwarf varieties. Coconut seedlings are planted in the centre of the pit. Two layers of coconut husk can be placed at the bottom of the pit before filling up for better moisture retention. The pits should be filled with two baskets of farm yard manure or compost and 100 gram neem oilcake per pit before planting.

Seedling can be transplanted with the onset of south-west monsoon during June-July. It can also be planted during October-November. Application of 50 g N, 40 g P_2O_5 and 135 g K_2O per plant is to be done after three months of planting. An adult plant is to be provided with 50 kg farm yard manure, 500 g N, 320 g P_2O_5 and 1200 g K_2O per year for better growth and yield. These are to be applied in two equal splits during June-July and October. Micronutrient application in the form of zinc and boron can enhance the yield. In case of acid soil, lime may be incorporated in the soil one month prior to application of chemical fertilizer. Application of 500 gram magnesium sulphate and 75 g borax per plant can be done during September-October. Application of organic matter enhances soil microbial activity and moisture retention capacity of the soil. It also increases aeration and water intake of soil. However, well decomposed organic matter is to be provided for better result.

Irrigation should be given adequately during the summer months. During kharif season, adequate drainage should be provided. The pits should be made weed free by hoeing and weeding. Burying fresh or dried coconut husks around the palm is a very beneficial practice for moisture retention in drought prone areas. The husks are to be placed in layers with concave surface facing upwards and covered with soil.

Major insect pests of coconut palm include rhinoceros beetle, red palm weevil, black headed caterpillar, eriophyid mite, etc. Rhinoceros beetle bores into the unopened fronds and spathes by making cuts. The beetles are to be taken out from the attacked plant by a hook. Spraying of profenophos in the breeding sites of the beetle can destroy the larva.

Red palm weevil causes holes on the stem, which results in oozing of viscous brown fluid through the hole. Sometimes, gnawing sound is produced by the grubs. This results in yellowing of the inner whorl of leaves. Clean cultivation, use of neem oilcake and avoiding injury to the trunk are some of the preventive measures against attack of insects. Injecting insecticide to the affected plant can control the insects. Black headed caterpillar feeds on green matter from lower surface of the leaf. Spraying of the plant with 0.02% profenofos can control this insect.

The eriophyid mite has become a major pest of coconut. It causes injury by feeding on the soft tissues. The symptoms are brown discolouration in patches of the husk. In case of severe attack, there is button shedding, which leads to poor yield. Application of 2% neem oil or neem based pesticides can control this insect. Besides, use of carbosulfan 25 EC @ 2 ml/ litre water can also control this insect.

Bud rot and stem bleeding are important diseases of coconut. The symptoms of bud rot are yellowing of younger leaves, black spots on leaves and dropping of crown. Application of Bordeaux paste on the affected part and spraying of carbendazim on the crown can control this disease. The symptom of stem bleeding is exudation of reddish brown liquid through cracks developed on the stem. The vigour and yield of the affected plant decline. It can be controlled by application of Calixin on the affected part after removing the infected tissue.

Fully developed nuts should be harvested one month before maturity. On an average, 80 to 100 nuts are obtained per tree/year. Green nuts are plucked at tender stage to get sweet water.

Papaya

Papaya is a highly nutritive fruit with medicinal values. The fruit is consumed as vegetable at unripe stage and as fruit at ripe stage. It is cheap and nutritious source of vegetable. It is a rich source of vitamin A and C. Papaya contains about 0.6 % protein, 0.1% fat, 7.2 % carbohydrate, 0.5 % minerals and 0.8 % fibre. As compared to other fruits, papaya produces higher yield per unit area. It is a tropical fruit and grows well in regions where summer temperature ranges from 35°C to 40°C. Papaya can tolerate frost and can be cultivated up to an elevation of 1200 meter above mean sea level. Well drained sandy loam soils are suitable for growing papaya. It is one of the most preferred plant in kitchen garden. It can be planted on the field or in containers/polybags placed on the rooftop. About three to four plants can meet the annual requirement of a family of four members. The promising varieties are Red Lady, Washington, Pusa Delicious, Coorg honeydew, CO1, CO2, CO 3, Surya, etc. Pusa Delicious, CO3, CO 7 and Surya are bisexual plants producing both male and female flowers on the same plant.

Papaya is propagated by seedlings raised from healthy seeds. Seedlings are raised in nursery beds or poly bags. About 500 g of seed is required for planting one hectare area. Fresh seeds are to be planted to get better germination. The seeds are to be treated with Thiram @ 3 g/kg of seeds. Seeds are dibbled in polybags at one cm depth. The polythene bags are kept in partial shade and watered with the help of rose can. Seedlings are ready for transplanting about 60 days after sowing. Seedlings can also be procured from reliable nurseries for planting in the home garden. The best time for planting is June-July. Two to three seedlings are planted in pits dug at a spacing of 1.8 m X 1.8 m. The length, breadth and depth of pit should be 45 cm each. The ratio of male and female plants should be 1:9 to get good fruiting. After flowering, the excess male plants are to be removed to maintain the desired sex ratio. This practice is not required for bisexual varieties.

Application of FYM @ 10 kg along with 20 g each of *Azospirillum* and Phosphobacterium per plant is done at the time of planting. Subsequently, 50 g each of N, P_2O_5 and K_2O per plant are applied at bi-monthly intervals from third month of planting. Spraying of zinc and boron during 4th and 8th month after planting can increase the yield. In home gardens, emphasis is given on application of organic manures such as farm yard manure, compost, vermi compost, oilcake, etc. to fulfil the nutrient requirement of the plant. The plants are irrigated once in a week or depending on soil moisture condition. Adequate drainage facility is to be provided during kharif season. The basin of the plant should be kept clean to avoid weed growth. Staking is necessary for tall plants to protect the plant from lodging due to high wind.

Yellow mosaic virus is the most important disease affecting papaya. It causes leaf mosaic and stunted growth of plant. A mottle or mosaic develops after 15-20 days of planting. The disease is caused by virus, which is transmitted by insect vector. Spraying of insecticides like fenvalarate, profenophos, etc. can control the vector. Anthracnose is another important disease affecting the plant. There is appearance of brown discoloured spots and slightly sunken areas on the fruits. Gradually the lesions coalesce and appear on the margins of the spot. This can be controlled by spraying of fungicides like carbendazim or propiconazole.

The fruit can be harvested either at raw or at ripe stage. Harvesting some fruits at raw stage as vegetable will induce more fruiting in the plant. In the kitchen garden, about 25-30 kg fruits can be harvested from one plant in a year. With good management and favourable climatic conditions, the yield can go up to 50 kg-60 kg per plant.

Guava

Guava is a nutritionally rich and delicious fruit crop. Guava contains about 0.9 % protein, 0.3 % fat, 11.2 % carbohydrate, 0.7 % minerals and 5.2 % fibre. It is a rich source of vitamin A, vitamin C, calcium, phosphorus and pectin. Guava gives handsome economic returns by involving little inputs. However, precise management techniques are to be adopted to get more profit. Guava can be planted in the backyard garden. It can also be planted in pots, containers, gunny bag or polybags placed on the rooftop or terrace. Various preserved products are prepared from guava. Guava can be grown both in wet and dry regions. It can be cultivated up to 1000 meter altitude. Well drained soil is suitable for guava cultivation. However, the best soils are deep and friable with pH ranging from 5.5 to 7.5. It can tolerate saline and alkaline condition.

Several varieties of guava are taken up by the farmers. Some of the varieties are Allahabad Safeda, Sardar, Lucknow 46, Arka Amulya, Arka Mridula, Banaras, Baptla, Arka Kiran, Lalit, etc.

Land preparation is done by deep ploughing followed by harrowing to kill the weeds and get good soil tilth. In home gardens, the soil should be tilled by spade and shovel to get a good soil tilth. Guava is propagated from seeds and seedlings. Trees from seeds are generally long-lived and produce fruits of variable size.

Layered plants produce fruits early and the fruits are uniform in size. Seedlings can be procured from established nurseries to get good yield. Planting is done in pits prepared at a spacing of 5-6 meter from plant to plant. The pits should be 45 cm in length, 45 cm in width and 45 cm in depth for better growth of the plant. Each pit is to be filled with 10 kg farm yard manure or compost and one kg neem or karanja oilcake. The seeds or seedlings are planted in the month of July to get better establishment. However, planting can be done during other months of the year with assured irrigation facility.

Application of 50 kg FYM or compost and one kg each of N, P_2O_5 and K_2O per plant may be done every year. It can be applied in two equal splits during the month of June-July and September-October. Spraying of 1 % urea along with 0.5% zinc sulphate twice a year can be done to obtain higher yield. In case of boron deficiency, spraying of 0.3% borax at flowering and fruit set stages is beneficial. In case of household plantation, emphasis is to be given for application of organic manures such as farm yard manure, compost, vermi-compost, oilcake, etc. to fulfil the nutrient requirement of the plant.

Hoeing may be done intermittently to kill the weeds. Weeds are to be removed from the container or pots for facilitating better plant growth. Irrigation may be done immediately after planting. Subsequently, irrigation can be done at 10-15 days interval as per need of the soil. There should not be water deficiency during flowering and fruiting stages. Adequate drainage facility may be provided during rainy season as the plant cannot withstand waterlogging condition. Pruning of old branches can be done during August-September or other months depending upon local climatic situation.

Tea mosquito is a common insect affecting guava. The symptoms are appearance of necrotic lesions on shoots and leaves. There is also appearance of rusty corky growth and scab on fruits. It can be controlled by spraying of Quinalphos 25 EC @2 ml per litre of water. Fruit fly is a harmful insect affecting guava plant. The fruits of affected plants are damaged severely. Spraying of profenophos @2 ml per litre of water can control this insect.

Die back is a serious fungal disease affecting guava. Due to this disease, the branches die from the top. Pin-head spots are seen on unripe fruits, which gradually enlarge. Fruit canker is another fungal disease, which affects the growth and yield of guava. The symptoms of fruit canker appear on the green fruits. There is appearance of small, brown or rusty necrotic patches on the fruits. These diseases can be controlled by application of carbendazim @ 2 g / litre water or propiconazole @ 2 ml /litre water.

Matured fruits can be plucked for harvesting. Harvesting may be done before ripening for enabling transportation to distant places. A fully grown plant yields about 100 kg fruits per plant per annum.

Sapota

Sapota, otherwise known as chiku, is cultivated for its nutritious fruits. Sapota is a delicious fruit mainly grown in humid tropical and subtropical regions. The fruits are round or ovoid, brown in colour and sweet in taste. Sapota contains dietary fibre, vitamin A, vitamin C, niacin, iron, potassium, copper and other minerals. It is a tropical plant and can be grown up to an altitude of 1000 metres above mean sea level. Sapota can be grown suitably with temperature range of 15-38^{0}C. High temperature adversely affects the flowering and fruit formation. It can be grown on wide range of soils ranging from sandy loam to clay loam. Some of the improved varieties of sapota are Kalipatti, CO-1, CO-2, PKM-5, Cricket Ball, Kirtibarathi, etc. In urban areas, sapota can be planted in the backyard. It can also be grown in pots, containers, polybags, etc.

Sapota is generally propagated though grafting and air layering. It is planted at a spacing of 8 m X 8 m in pits of 1 m x 1 m x 1 m size. The planting distance can be changed depending on the variety and soil type. It is planted in the month of July-August. Pits are to be filled up with 10 kg FYM or compost and one kg neem cake. The plants are to be staked with bamboo or wooden stick to avoid damage of the graft joints by heavy wind. An adult plant is to be provided with 50 kg of FYM or compost, one kg N, one kg P_2O_5 and 1.5 kg K_2O every year. Manures and fertilizers are to be applied in two equal splits during June-July and October. The time of fertilizer application varies depending on monsoon rain and local climatic condition. In home gardens, organic matter is preferred instead of chemical fertilizer.

Hoeing may be done intermittently to kill the weeds. The containers with sapota plant should be thoroughly cleaned for better plant growth. Sapota can tolerate scarcity of soil moisture to some extent. For good growth and high yield, plants may be irrigated during summer and winter seasons. Irrigation may be done immediately after planting. Subsequent irrigations can be provided at an interval of 7- 10 days during summer and 20-25 days during winter season. There should not be water deficiency during flowering and fruiting. Adequate drainage facility may be provided during rainy season. Removal of the rootstock sprouts, water shoots and lower branches helps to obtain higher yield. Training of the plant is done by removing the lower branches up to 90 cm above the ground.

Sapota is affected by the insect chikoo moth. The larvae of the insect feed on leaves, buds, flowers and tender fruits. Green scale insect causes serious loss to the plant. The nymphs and adults suck the sap from leaves, which results in yellowing of leaves. Spraying of profenophos or dimethoate @ 2ml per litre of water can control these insects.

Fruits are harvested when those are fully matured and dull brown in colour. The mature fruits shed off brown scaly scurf from the skin. When matured, fruits do not show green tissue underneath the brown skin. The fruits are harvested by hand picking. Depending upon the variety and management practices, an adult plant can produce about 700 fruits per year.

Pomegranate

Pomegranate is one of the most favourite fruits used for table purpose. Pomegranate requires minimum care and maintenance for growing. The fruits are delicious, red in colour with leathery rind. Pomegranate is nutritious and rich in minerals, vitamins and proteins. Mature fruits contain about 1.6 % protein, 0.1% fat, 14.5 % carbohydrate, 0.7 % minerals and 5.1 % fibre. The seeds are full of antioxidants and have several health benefits. It is grown in a wide range of soils and can tolerant salinity & alkalinity. Soil having pH range from 6.5 to 7.5 is ideal for pomegranate cultivation. Pomegranate can tolerate water stress condition. Cool winter and dry summer are necessary for production of good quality fruits. It can be grown up to an altitude of 1800 m above mean sea level. Some of the promising varieties are Jyothi, Ganesh, Co 1, YCD 1, Araktha, Rudhra, Ruby, Mirudhula, etc. It can be planted near the house premises with minimum available space. However, regular pruning is necessary to restrict the growth of the foliage.

Pomegranate is propagated by 12 to 18 months old rooted cuttings or layer. The ideal time of planting is July to August. Planting is done in pits of 60 cm x 60 cm x 60 cm size at a spacing of 3 m X 3 m. At the time of planting, ten kg of FYM and one kg neem/karanja oil cake is applied per pit. After establishment of the plant, fertilizer dose of 200 g N, 100 g P_2O_5 and 400 g K_2O is to be applied to each plant. Subsequently, the fertilizer dose is enhanced with advancement of age of the plant. An adult plant of 6 years and above may be provided with 30 kg FYM, 600 g N, 500 g P_2O_5 and 1200 g K_2O per year. The fertilizers are to be applied in two equal splits during June-July and October. In household plantation; organic matters like farm yard manure, compost, vermi compost, oilcake, etc. are to be applied to get good quality fruits. Annual pruning is done after harvest of fruits to promote new shoot growth. The dried and diseased branches are to be removed. Irrigation is required during summer and winter months as per soil moisture status.

Fruit borer is the major insect affecting pomegranate plant. Pomegranate fruit borer is the most widespread, polyphagous and destructive pest. It is more prevalent during monsoon season. Infestation occurs from flowering to button stage causing severe yield loss. The larvae of the insect bore into the fruits and plug the hole by the abdominal segment. The fruits rot and drop off due to infestation of this insect. The fruits are to be bagged with polythene covers during flowering period to prevent egg-laying of the insect. Spraying of neem oil or neem based pesticide can control this insect. Two sprays of emamectin benzoate 5 SG @ 0.25 g/lit water or spinosad 45 SC @ 0.20 ml/lit water can control the insect. Besides, application of neem oil or neem based insecticides can also control this insect.

Pomegranate fruits are harvested 3-4 years after planting. The fruits are harvested by carefully cutting from the branch. Depending upon the variety and management practices, a fully grown tree can produce about 12 to 15 kg fruits per year.

Sweet Orange

Sweet orange is one of the most important citrus species grown in India. It has several health benefits and has medicinal properties. Ripe orange contains about 0.7 % protein, 0.2 % fat, 0.3 % minerals and 0.3 % fibre. It is a good source of vitamin C, potassium and minerals. It can be grown on the backyard or in the containers placed on rooftop. Well drained loamy soil with pH ranging from 6.5 to 7.5 is suitable for sweet orange cultivation. This plant does not perform well in waterlogged condition. It can be grown at an elevation up to 500 m above mean sea level. Several varieties are available for cultivation of sweet orange. Nagpur Santra, Coorg Santra, Khasi Santra, Mudkhed, etc. are some of the popular varieties. August-September is the ideal time for planting of sweet orange. Pits are dug at a distance of 7m x 7 m for plantation of sweet orange. The size of the pit should be 75 cm x 75 cm x 75 cm for better root growth. The pits are to be filled up with 10 kg well decomposed farm yard manure or compost, one kg neem/karanja oilcake and top soil before planting. An adult plant is provided with 30 kg well decomposed farm yard manure or compost, 600 g N, 200 g P_2O_5 and 300 g K_2O per year. The fertilizers are to be applied in two equal splits during June-July and October. In home gardens; organic matters like farm yard manure, compost, vermi compost, oilcake, etc. are to be applied to get good quality fruits. Manuring is done in basins prepared about 70 cm away from the plant. Intermittent hoeing is necessary for controlling the weeds and loosening the soil.

First irrigation is given immediately after planting. Subsequent irrigations are provided at an interval of 10 days depending on soil moisture status. Water stagnation is to be avoided near the plant during monsoon season. Application of micronutrients such as zinc and boron helps in better plant growth. Pruning is necessary for obtaining higher yield. The diseased, damaged or dropped branches are to be removed periodically to obtain maximum vigour. Pruning may be avoided when the plants are at bearing stage. Fruit dropping is a major problem of sweet orange. Spraying of 2,4-D at 10 ppm at the time of flowering can control fruit dropping.

Citrus canker is an important disease of sweet orange caused by bacterium. This disease causes premature leaf falling & fruit drop, twig dieback and deformation of fruits. Application of plantomycin along with carbendazim or mancozeb can control this disease. Dieback is a serious disease affecting orange plant. The symptoms of this disease appear on older leaves. The mid rib and veins become yellow and later change into orange yellow. Plant growth is stunted and gradually the branches die from the tip. This can be controlled by application of fungicides like propiconazole or cupper oxychloride. Sometimes, the plant is affected by leaf miner. The plant growth is adversely affected due to this insect. It can be controlled by spraying profenophos or dimethoate@ 2 ml/litre of water.

Harvesting of sweet orange starts about five years after planting. Fruits are plucked at mature stage for harvesting. An adult plant can yield about 150-200 fruits per year depending upon the variety and management practices.

Ber

Ber or Jujube is one of the most ancient cultivated fruit tree of India. It can be grown in arid and semi-arid areas. The fruits are rich in vitamins and various minerals. It contains about 0.8 % protein, 0.3 % fat, 17.0 % carbohydrate and 0.3 % minerals. The thorny branches are used as fencing material. *Ber* is a hardy tree adapted for growing under diversified agro-climatic conditions. It can tolerate high temperature, but fruit setting is adversely affected at temperature above 35°C. It can be grown in marginal land having minimum fertility status. This tree can be taken up in a wide range of soils. Neutral or slightly alkaline soils are suitable for this plant. This can be taken up in the backyard or in the pots placed on the rooftop or terrace. As the plant is having thorns, it can be planted near the boundary or fence to protect the home garden. There are several varieties of *ber* suitable for cultivation in different areas. Some of those are Gala, Mundia, Banarasl, Umran, Maharwali, Kaithli, etc.

Ber can be propagated by seeds or budded plants. Budded plants help in obtaining uniform size and good quality fruits. Normally, planting is done in the pits during June-July. The pits are prepared at a distance of 6-7 m for growing ber. The pits should be of size 60 cm x60 cm x60 cm. Each pit is to be filled up with a mixture of top soil, 10-15 kg FYM and one kg neem cake. Planting is done at the middle of the pit. After planting, the soil around the plant needs to be pressed. Watering should be done immediately for better establishment. In some areas, in-situ budding is also done in the main field.

Requirement of nutrients should be decided after analysis of soil nutrient status. Generally, a full grown tree is provided with 50 kg farmyard manure, 500 g N, 200 g P_2O_5 and 500 g K_2O per year. Fertilizer is to be applied in two equal splits during June-July and October. However, application of fertilizer immediately after pruning is helpful. Water soluble fertilizer can also be applied to the plants grown in pots or containers to avoid scorching effect. Foliar application of boron and zinc twice during flowering and fruit setting increases yield and quality. In household plantation; organic matters like farm yard manure, compost, vermicompost, oilcake, etc. are to be applied to get good quality fruits.

Hoeing should be done around the plant to make the area weed free. Hoeing and earthing up around the tree kills the hibernating insect, disease causing pathogen and weed. Pruning of old branches is necessary to enhance fruit production. Pruning of side branches is done every year to obtain good yield.

Ber plant is affected by various diseases. Powdery mildew is one such fungal disease causing severe loss to the plant. The symptoms are seen on the leaves and newly developed fruits. Young leaves show whitish powdery mass, which result in leaf shrivelling and defoliation. The disease can be controlled by spraying sulfex. In case of recurring occurrence of the disease, a prophylactic spray of fungicide like carbendazim or propiconazole may be done on the plant after pruning. Fruit fly is a serious insect affecting the plant growth and yield. The infestation starts at tender stage of the fruits. The larvae feed on the young fruits and make those unfit for consumption. Spraying of Methyl Demeton or Profenophos@ 2 ml per litre water can control this insect.

Harvesting should be done at right stage of maturity to obtain good quality fruits. The fruits are to be harvested when the skin colour becomes golden yellow. Premature or over mature fruits are not preferred in the market. Fruiting should be allowed from 3-4 years after planting. The peak harvest of fruit reaches at 8-10 years after planting. Under rainfed situation, the yield per plant is about 50-60 kg per annum.

Stone Apple (*Bael*)

Stone apple or *bael* is a fruit tree having medicinal value and religious importance. *Bael* leaf has enormous medicinal value to cure various stomach related diseases. Now-a-days, people prefer planting *bael* tree in the backyard to fulfil daily requirement of the leaves. Sandy loam soil with pH ranging from 5.0 to 8.0 is suitable for *bael* planting. Warm humid climate receiving abundant sunlight is suitable for cultivation of this plant. There are many varieties of *bael* to obtain good yield. Some of the examples are Narendra Bael 1, Narendra Bael 2, Narendra Bael 5, CISH B-1, CISH B-2, Goma Yashi, Pant Aparna, Pant Shivani, Pant Urvashi, etc.

Bael is propagated by seed, grafts or layered plants. The seeds are to be soaked in water for about 12-14 hours before planting. Seeds are planted in polybags or raised beds for preparation of seedlings. The field should be cleaned for planting the tree. Pits are made at a distance of 8-10 meter. The size of the pit should be 90 cm x 90 cm x 90 cm. The pits are to be filled up with 25 kg FYM or compost and one kg of *neem/karanja* oilcake before planting. *Bael* is planted in pits during July-August. Water is to be provided immediately after planting for better plant establishment. After one year of planting, application of 10 kg farmyard manure, 50 gm nitrogen, 25 gm phosphorus and 50 gm potassium is to be done. The quantity of fertilizer is to be enhanced every year as per the growth of the plant. For a fully grown plant, application of 10 kg farmyard manure or compost, 500 gm nitrogen, 250 gm phosphorus and 500 gm potassium is to be done per annum to get better growth and yield. Fertilizer is to be applied in two equal splits during the months of June-July and October. Organic matters like farm yard manure, compost, vermi compost, oilcake, etc. are to be applied to get good quality fruits. Pruning is necessary to give proper shape to the plants and obtain higher fruit yield.

Weeds should be removed from the field. Hoeing and weeding is to be done intermittently to control the weeds. Based on soil moisture status, the plant is to be watered in summer and winter months. Adequate drainage is to be provided during rainy season.

Bael plant is affected by various insects and diseases. Fruit canker is an important disease affecting the plant. Due to this disease, spots are observed on leaves and fruits. Affected twigs and branches are to be removed to prevent spread of the disease. Application of streptomycin sulphate @ 2 g/10 litre water along with copper oxychloride @ 3 g/litre water can control this disease. Leaf eating caterpillar is a major insect causing harm to this tree. It damages the young leaves and adversely affects plant growth. It can be controlled by spraying of profenophos @ 2 ml per litre of water.

The plant starts producing fruits about 6 to 7 years after planting. Harvesting is done when fruits become yellowish-green in colour. A fully grown tree can produce about 150-200 fruits per year. However, it varies depending on variety, climatic condition and management practices.

Pineapple

Pineapple is a commercially important fruit crop grown in India. It is a good source of vitamin C, fibre and minerals such as calcium, magnesium, iron etc. Ripe fruits contain about 0.6 % protein, 0.3% fat, 12.0 % carbohydrate, 0.5 % minerals and 0.5 % fibre. Mild tropical climate with adequate humidity is suitable for this crop. This crop can also be taken up under partial shade. In homestead area, pineapple can be planted near the fence or under shade condition. It can be cultivated up to an elevation of 500-700 m above mean sea level. Well drained light soil with pH ranging from 5.5 to 7.0 is suitable for this crop. It can also be cultivated in heavy soil with adequate drainage facility. Several varieties are cultivated in the country, which include Kew, Giant kew, Mauritius, Queen, etc.

Pineapple is normally grown during rainy season. It is planted in the month of July-August. Suckers and slips of 300-350 g weight can be used as planting material. The suckers are to be cut in a slanting manner for planting. The suckers are dipped in Mancozeb or Carbendazim to prevent spread of fungal disease. Planting is done in double rows either in beds or in trenches. The spacing between two trenches is 90 cm and row to row spacing is 60 cm. The plant to plant spacing within the row is maintained at 30 cm. Application of FYM or compost @ 40-50 kg/10m^2 area is to be done at the time of final land preparation. Application of 16 g N, 4 g P_2O_5 and 12 g K_2O per plant is to be done in two equal splits at 6 and 12 months after planting. Foliar spray of zinc and boron can enhance the yield. Spraying of 200-300 ppm NAA after fruit formation increases the size of the fruit. Irrigation may be provided during summer and winter season as per need of the soil.

Among insects, mealy bug causes substantial harm to the crop. The insect sucks the sap for which the plant growth is adversely affected. It can be controlled by spraying of methyl demeton or profenophos @ 2 ml/litre of water. Phytophthora root rot is a fungal disease that affects the growth and yield of pineapple. Due to this disease, the leaf tips and margins become necrotic. The roots die and the plants are easily pulled from the ground. The disease can be controlled by application of copper oxychloride or mancozeb.

Pineapple is ready to harvest at about 18-24 months after planting. It is harvested when there is slight change of colour at the base of the fruit. The yield is about

50 tonne per hectare. In case of home garden, about 50 kg pineapple can be obtained from 10 m^2 area.

Lemon

Lemon is one of the most important fruit used in daily diet of every household. Lemon is rich in vitamin C along with other compounds beneficial for human health. Lemons can be grown in all types of soil. However, light textured soils with good drainage are most suitable. The ideal soil pH ranges from 5.5 to 7.5. Lemon can tolerate slightly alkaline and acidic soil reaction. The crop can be cultivated in both tropical and subtropical climates. It can be grown up to 1000 meter above MSL. In urban areas, one or two plants of lemon in the kitchen garden can fulfil the family requirement. In some cases, lemon plant is grown in the pot, container or polybag and placed on the rooftop or terrace. There are different types of lemon such as Assam Nimbu, Lisbon, Genoa, Rough Lemon, Nepali Oblong, Pant Lemon 1, Rasraj, Villafranca, Eureka, Vikram, Lucknow Seedless, Kagzi Kalan, Nepali Round, etc.

Planting of lemon requires thorough land preparation. The land should be prepared properly to get a clod free and clean field. In backyard, the land can be cleaned by spade before planting of lemon. Ideal time of planting is July-August. Lemon plants are propagated by budding or air layering. Healthy and disease free seedlings are planted at a distance of 5.0 m X 5.0 m in the pits of 60 cm X 60 cm X 60 cm size. The pits are to be filled with 10 kg of FYM and 80 g of phosphorus before planting. After one year, 10 kg FYM, 200 g N and 100 g each of P_2O_5 & K_2O are to be applied per plant for better growth. The fertilizer dose is increased every year depending on the growth of the plant. From 6th year onwards, 30 kg FYM, 600 g N, 200 g P_2O_5 and 300g K_2O are to be applied per plant to get good yield. Spraying of Zinc sulphate @ 5 g/ litre water after emergence of new flushes will increase the number of fruits per plant. In home gardens; organic matters like farm yard manure, compost, vermi-compost, oilcake, etc. are to be applied to get good quality fruits.

The area adjacent to the plants should be kept weed free by hand weeding/ hoeing. After establishment of the plant, irrigation is given as per soil moisture condition. Water stagnation should be avoided during rainy season. Side shoots should be removed to give the plant proper shape. Water suckers are to be removed at early stages of growth. Spraying of 2, 4–D @ 20 ppm during flowering stage enhances fruiting.

Lemon plant is affected by various insects and diseases. Leaf miner is one of the serious pest of lemon. The symptoms are curling of young and newly emerged leaves. Young plants show reduction in growth due to this insect attack. It can be controlled by spraying of Profenophos @ 2.0 ml per litre of water. Pheromone traps are also used to control the pest. Alternatively, neem seed kernel extract or neem oil can be sprayed to control the insect.

Citrus canker is an important disease found in lemon. The affected plants develop brown lesions with water-soaked margins on stems, leaves and fruits. Younger leaves are highly susceptible to this disease. It can be controlled by spraying of streptomycin sulphate along with propiconazole. Lemon scab is a fungal disease, which affects the plant. Grey coloured scabs are seen on branches, leaves and fruits resulting yield loss. There is fruit dropping at early stages of growth. Spraying of propiconazole @ 2 ml/lit. or carbendazim @ 2 g/lit. or mancozeb @ 3 g/litre water can control this disease.

The plant starts bearing fruits from third year after planting. A single plant yields about 50-60 fruits per year during initial fruiting stage. A fully developed plant can yield up to 500-600 fruits per year harvested in two to three phases. However, yield of fruit depends on the variety, climatic condition and management practices.

Litchi

Litchi is a delicious fruit popularly grown in India for its soft edible flesh. The fruit contains sugar, vitamin B1, Riboflavin, vitamin C, carbohydrate and minerals like calcium, phosphorus, iron, etc. Litchi is used for preparation of excellent canned fruit. Many processed products such as juice, squash, pickles, etc. are prepared from litchi. It thrives well under moist sub-tropical climate. Litchi can be grown up to an altitude of 800 m above mean sea level. Deep, well drained loamy soil, rich in organic matter with pH ranging from 5.0 to 7.0 is ideal for litchi plantation. Litchi cannot tolerate frost or very high temperature. Temperature above 40.5 ^{0}C during summer season is harmful for this plant. Prolonged rain at the time of flowering may be detrimental for litchi production. In home gardens, one or two trees can be planted in the backyard.

China, Deshi, Purbi, Bedana, Muzzaffarpur, Rose Scented, Shahi, Kasba, Bombai, Saharanpur, Dehradun, Calcuttia, Seedless, Early Large Red, Late Large Red, Elachi Early, Elachi Late, etc. are some of the cultivated varieties of litchi. Air layering is the most common method of propagation of litchi. It is usually done at the beginning of monsoon season.

As litchi plant is susceptible to damage by high wind, it should be planted near boundary wall or tall building in the home gardens. Tall growing trees such as mango, jamun, eucalyptus, deodar, etc. can also protect litchi plant from high wind. Planting is usually done during July-August. However, with assured irrigation facility, planting can be done throughout the year. Planting is done in pits of 1m X 1m X 1m size dug at a distance of 8-10 meter from plant to plant. Each pit is filled with 20-25 kg FYM, 2 kg bone meal and 300 g muriate of potash before panting. The seedlings are planted at the centre of the pit and watering is done immediately after planting.

Fertilizer application can be done as per soil test results. However, an adult plant requires 60 kg FYM, 700 g nitrogen, 350 g phosphorus and 350 g potassium per year. The fertilizer is to be applied in two equal splits during the months of June-July and October. The plants are to be irrigated during summer and winter months as per requirement of the soil. Adequate drainage is to be provided during rainy season for better plant growth. The weeds around the plant need to be controlled by hoeing or other means.

Pruning is necessary to provide proper shape to the plant. Dead or diseased branches and damaged shoots are to be removed. In case of too much vegetative growth, roots and shoots need to be pruned to give good shape to the plant. Heavy pruning is required when the trees become old.

Several insects and pests affect litchi plant. Shoot borer is a common insect attacking the plant during rainy season. The insect enters into the stem and causes damage to the plant. It can be controlled by spraying fenvalarate or methyl demeton @ 2 ml per litre water. Sometimes, the plant is affected by litchi mite. The mites puncture the leaf and suck the cell sap. There is appearance of hairy blister like gall on the upper side of young leaves. The leaves become wrinkled, distorted and ultimately fall off. Spraying of neem oil 5% or neem based pesticide is advisable for control of the mites.

Fruiting begins at about 5- 6 years after planting. After flowering, it takes 70 to 100 days for the fruits to ripe. Maturity of the fruit is judged by change of fruit colour from green to pink. The fruits are harvested in bunches along with a portion of the branch and a few leaves. Depending on variety, local condition and management practices, a litchi tree can yield about 50-60 kg fruits per annum.

Custard Apple

Custard apple is also known as Sitaphala. It is a hardy plant and can be grown under various adverse situations. The fruits are consumed after ripening. Various products like custard powder and ice-cream are prepared from the fruits. Unripe fruits, seeds, leaves and roots are used for medicinal preparations. This plant can be grown in urban areas in comparatively less fertile land. As custard apple requires minimum care, it can be taken up in any unused land in the homestead area. Custard apple requires hot and dry climate during flowering and high humidity at the time of fruit setting. Low humidity is harmful for pollination and fertilization. The plant can withstand water stress condition to some extent. Custard apple can be cultivated in various types of soils like shallow, sandy, red loamy, etc. It can also be grown in deep black soil with proper drainage.

Several varieties of custard apple are cultivated for higher yield. Some of the varieties grown in different agro-climatic regions are Red Sitaphal, Balanagar, Washington, Purandhar, etc. Custard apple is mainly propagated by seeds. However, vegetative methods can be adopted for multiplication of the plant.

Planting is done during the months of July-August. For the purpose, pits are dug at a distance of 5-6 meter. The size of the pit is 60 cm X 60 cm X 60 cm. The pits are filled with well decomposed FYM, single super phosphate and neem or karanja oilcake. Application of manure and fertilizer at the onset of monsoon and at cessation of monsoon will increase the fruit yield. Weeding should be done to keep the field clean. Application of 10-20 ppm NAA just before flowering can reduce the problem of fruit dropping. Custard apple is grown under rainfed situation. But irrigation may be provided during summer season to obtain good yield. Water stagnation during monsoon season adversely affects plant growth.

Being a hardy plant, there is minimum attack of insect and disease. However, in some cases, mealy bug and scale insects attack the plant. The plant is also affected by leaf spot and anthracnose diseases. Spraying with neem oil or neem based pesticide can control the insect and disease.

Custard apple is harvested when the fruit changes its colour from green to various colour shades. Fruits are harvested by plucking with hand. A properly managed tree can yield about 100 fruits per annum. The weight of individual fruit varies between 300 to 400 grams. Normally, fruits are harvested during the month of August to October.

Jackfruit

Jackfruit is one of the most important fruits found in India. Jackfruit is found in wild form in various parts of the country. Now-a-days, it is planted commercially for making profit. The fruits are used for culinary purpose at tender stage. The ripen fruits are used for table purpose. As vegetable, jackfruit enjoys a high demand and premier price. Especially in urban markets, jackfruit is sold at a higher price. It can be planted in any unused land either in the homestead garden or in public place. Ripe fruits are very sweet and have high nutritive value. The fruit contains minerals and various vitamins. The fruits are also used for preparation of pickles, dehydrated leather, thin papad, squash, etc. The skin of the fruit and leaves are used as cattle feed. Jackfruit tree also provides valuable timber for making furniture. Jackfruit requires fertile and well drained sandy loam soil. This tree cannot tolerate moisture stress condition. Areas near the river beds are ideally suitable for jackfruit cultivation. Warm humid climate is suitable for jackfruit cultivation. It can be grown up to an elevation of 1,500 meters above mean sea level. Cold weather and frost adversely affect growth of this plant. Normally, wild species are found in various parts of the country. The fruits vary in shape, size, rind colour, fibre content, fruit quality and maturity. Cultivated types are broadly classified into two groups viz. soft flesh and firm flesh. Gulabi, Champa, Hazaric, etc. are some of the local types. Some of the improved varieties are Black Gold, Cheena, Dang Rasimi, Golden Nugget, Golden Pillow, etc.

Jackfruit is generally propagated by seeds, which results in wide variation in fruit quality. The seeds are sown either in beds, polythene bags or earthen pots. Well developed, healthier seeds are used for sowing. The seeds are soaked for 24 hours in water before sowing. It helps in better germination and seedling growth. Efforts are on for standardization of propagation by vegetative means. Inarching and layering have been found effective in multiplying the plant. Planting is done in the pits with one meter each of length, breadth and depth. Planting is done at a distance of 12 meter from plant to plant. About 30 kg well decomposed farmyard manure and 500 g super phosphate are applied to each pit before planting. Three to four seeds can be planted per pit to ensure desired plant stand. After planting, the soil is pressed thoroughly for better plant establishment. The ideal time for planting of jackfruit is June to August. Application of 30 kg FYM, 75 g N, 60 g P_2O_5 and 50 g K_2O is to be done per plant per year up to 8 years after planting. Subsequently, application of 30 kg FYM, 200 g N, 100 g P_2O_5 and 100 g K_2O is recommended per plant per year. Fertilizer is to be applied in two equal splits during June-July and October. The young plants need to be protected from stray animals. Watering of young plant is necessary during summer season. Regular weeding is to be done to achieve normal plant growth.

Several insects and diseases affect jackfruit plant. The shoot and trunk borer enters into the tender shoots and buds. As a result, the affected plant parts dry and ultimately the buds start dropping. Application of methyle demeton or profenophos can control this insect. Soft rot is common disease affecting jackfruit. It is a fungal disease badly affecting the male flowers and small fruits. Humid atmosphere facilitates spread of this disease. Spraying of carbendazim or copper oxychloride can control the disease infestation.

Jackfruit starts bearing fruits after 7-8 year of plantation. The pick harvest is obtained about 15-16 years after planting. Tender fruits are harvested for use as vegetables. Harvesting is done by cutting the stalks carrying the fruits. Normally, a tree bears 50-100 fruits annually. The yield of the fruit varies depending on the variety, climatic condition and management practices.

6

Vegetable Crops

Vegetable refers to the edible parts of plant such as leaf, root, stem, fruit, rhizome, corm or seeds consumed directly or after cooking. Vegetables are an important constituent of balanced diet taken for good health. Vegetables are good source of dietary fibre, protein, essential vitamins, minerals and antioxidants. Hence, vegetables play a vital role for human health. In urban areas, vegetables can be grown in the backyard, rooftop and terrace. Harvesting vegetables from own garden has a separate satisfaction. The vegetables produced in home gardens are of superior quality as those do not contain any harmful toxic substance. Growing vegetables in the home premises will reduce the daily expenditure of a family. In urban areas, vegetable can be taken in small patches in the backyard to fulfil the family requirement. In absence of surplus land area, it can be grown in pots, containers or polythene bags placed on the rooftop or terrace. Some vegetables can also be grown on the rooftop in a specially prepared bed with shallow depth of soil. In this chapter, discussion is made for cultivation of vegetables in small patches.

Tomato

Tomato is an important vegetable crop grown across India. It is a nutritious vegetable containing adequate quantity of protein, carbohydrate, vitamin A, vitamin C and minerals. This can be grown both on the bed and in containers or pots. It is primarily a *rabi* season crop, planted during the months of October- November. In hilly areas and cooler plains, it can be grown during summer and *kharif* seasons. The optimum temperature requirement for this crop is 15-27^{0} C. Temperature beyond this range may cause poor germination, retarded plant growth, flower drop and low fruit setting. There are several high yielding varieties such as Punjab Chhuara, Pusa Ruby, Vaishali, Arka Alok, S-152, Roma, Co 1, Pusa Early Dwarf, Arka Vikas, Arka Saurabh, Arka Meghali, Arka Ahuti, Arka Ashish, Arka Abha, Arka Shreshta, Arka Abhijit, etc. cultivated in various parts of the country. Some of the promising hybrids of tomato are Swarna Sampad, Megha, Rajni, Pusa Bahar, etc.

Tomato requires well drained loamy soil with adequate organic matter. The soil pH should be in the range of 6.5-7.5 for growing tomato. About 0.5 gram seed is required to grow tomato in 10 m^2 area. Seeds are to be treated with *Trichoderma viride* or *Pseudomonas fluorescens* before sowing. Alternatively, the seeds can be treated with Carbendazim or thiram to prevent attack of seed borne diseases. Seedlings are grown in raised nursery beds. Application of FYM, neem cake, VAM and single super phosphate in the nursery bed will produce healthy seedling. The seeds are to be sown in raised nursery beds with a line spacing of 10 cm.

Tomato requires a fine soil tilth. Hence, the land is to be prepared thoroughly to obtain good tilth. In backyard area, land can be prepared by digging the land with the help of spade or other implement. FYM is added @ 25 kg per 10 m^2 area before planting. Application of neem or karanja oilcake @ 0.25 kg per 10 m^2 area is helpful for obtaining healthy plant. Ridges and furrows are made at a spacing of 60-75 cm for transplanting of seedlings. *Azospirillum* and Phosphobacteria are to be applied by mixing with FYM for accelerated plant growth. Twenty-five days old seedlings are to be transplanted at a plant to plant spacing of 30 cm. If the soil is dry, the furrows are to be irrigated prior to transplanting. Light irrigation is required two to three days after planting. Seedlings can also be planted in the pots containing suitable pot mixture. Equal quantities of sand, loamy soil and farm yard manure are taken for preparation of pot mixture. Leaf mould can also be added in the pot mixture to get better result. In case of planting in pots, adequate care is necessary for drainage of excess water.

Tomato requires adequate quantity of nutrition for plant growth and fruiting. Fertilizer may be applied based on soil test results. However, for high yielding varieties, 120 gram nitrogen 600 gram phosphorus and 50 gram potassium is applied in 10 m^2 area. Hybrids require higher quantity of fertilizer for better yield. All the phosphorous, 50% of potassium and 25% of nitrogen is applied as basal dose. Application of Borax @ 10g/10 m^2 and Zinc sulphate @ 50 g / 10 m^2 is to be done to address micronutrient deficiency. Topdressing of 50 % of nitrogen is to be done at 15-30 days after planting. Balance quantity of nitrogen and potassium are applied at 45 to 60 days after planting. Boron can also be sprayed in the standing crop for better plant growth. Staking of the plants may be done with 1-1.5 m tall sticks at 30 days after planting. Judicious watering helps in better plant growth and yield. The critical periods of water requirement in tomato are initial establishment, flowering and fruit setting. Irrigation water is regulated in such a manner so that two third of the furrows are filled up. Water stagnation, either in the pot or in the field is detrimental for plant growth.

Tomato suffers due to attack of several diseases like bacterial wilt, leaf curl, damping off, etc. Bacterial wilt is one of the most serious diseases of tomato. The affected plants wilt rapidly and die. Sometimes, the lower leaves may drop before wilting. Application of plantomycin and copper oxychloride can control this disease. Movement of water from affected plots should not be allowed to enter adjacent plots to check the spread of this disease.

In case of leaf curl disease, there is curling of leaves and reduction in internode length. Flowers may drop due to the disease infestation. This disease is spread by insect vectors. Hence, control of insects can be done by spraying of imidachloprid or dimethoate at 15, 25 and 45 days after transplanting. This will check the spread of the disease. Damping off is another disease, which occurs at the early seedling stage. The infected seedlings die just after emergence due to fungal attack. This can be controlled by application of 0.2% Metalaxyl or soil drenching with copper oxychloride 0.2%. Seed treatment with *Trichoderma viride* can prevent incidence of the disease.

Tomato is affected by insects like fruit borer, white fly, etc. Fruit borer feeds on tender foliage and enter the fruits. The affected fruits get spoiled due to borer attack. This can be controlled by release of *Trichogramma pretiosum* @ 100/ 10 m^2 at an interval of 7 days starting from flower initiation. Spraying of neem based pesticide or profenophos can also control this insect. Tomato is also affected by whitefly. The symptoms of whitefly attack are chlorotic spots, yellowing of plant and drying of leaves. This insect also acts as a vector for various viral diseases affecting the plant. Application of Dimethoate 30 % EC or Thiamethoxam 25 % WG can control this insect.

A good crop of tomato can yield about 30-40 kg fruits per 10 m^2 area. In case of hybrids, the yield can be higher. In case of pot culture, about 10 plants can fulfil the requirement of a small family.

Brinjal

Brinjal is a popular vegetable crop grown in most of the states of India. It can adopt to various climatic situations prevailing across the country. It is mostly used for culinary purpose. A large number of cultivars of various size, shape and colour of fruits are grown in India. Brinjal is a moderate source of vitamins and minerals like phosphorous, calcium and iron. Brinjal can be grown in pots, containers or polythene bags. Besides, it can be planted on the ground in home garden. There are several high yielding varieties and hybrids available in the market. Some of the promising high yielding varieties are Pusa Purple Long, Pusa Purple Cluster, Pusa Kranti, Pusa Purple Round, Pusa Shymala, Azad Kranti, Arka Shirish, Utkal Anushree, etc. Some of the hybrids are Pusa

Hybrid 5, Pusa Hybrid 6, Pusa Hybrid 9, Pusa Anmol, Arka Navaneeth, Pant Brinjal Hybid- 1, Phule Hybrid 1, etc.

Brinjal can be suitably grown in well drained fertile loamy soil. The ideal pH range for cultivation of brinjal is 6.5-7.5. This can be grown round the year, but performs better during *kharif* and summer seasons. The *kharif* crop is planted during June-July and summer crop can be planted during January. However, time of planting may vary depending upon local climatic conditions. It can be grown with a mean temperature range varying from 20^0 C to 30^0 C. Severe frost or low night temperature is detrimental for this crop.

Brinjal is planted after raising of seedlings. The seedlings are prepared in raised nursery beds at a height of 15 cm above the ground level. Application of FYM, neem oilcake, VAM and single super phosphate is to be done in nursery bed to get healthy seedlings. About 0.5 g seed is required for planting in an area of 10 m^2. Less quantity of seed is required in case of growing hybrids. Seeds are treated with *Trichoderma viride* or *Pseudomonas fluorescens*. The seeds are sown in nursery beds at a row to row distance of 10 cm. The seedlings are transplanted after 3-4 weeks of sowing at a spacing 60 cm X 45 cm in the backyard. The seedlings can also be transplanted in the pots or container. Before transplanting of the seedlings, the field is to be thoroughly ploughed to make fine tilth. Some plots can be prepared with the help of spade and other implements. Application of 25 kg farm yard manure or compost along with 250 gram neem or karanja oilcake is to be done in 10 m^2 area before planting. Brinjal requires 100-120 g nitrogen, 60-80 g P_2O_5 and 50-60 g K_2O in an area of 10 m^2. All of P_2O_5 and K_2O along with 25 % N are applied as basal dose. Rest of nitrogen is applied in two equal splits at 20-25 and 40-45 days after planting. However, fertilizer may be applied based on soil test results. Use of water soluble fertilizer is preferred in home gardens to avoid scorching of the plants. Application of bio-fertilizer like *Azospirillum* and Phosphobacteria during transplanting is beneficial. Foliar spray of boron and zinc helps in better plant growth and fruit setting. In home gardens, emphasis is given on application of organic manures like farm yard manure, compost, vermi compost, oilcake, etc. to provide plant nutrients. In case of potted plants, the potting mixture is prepared by using equal quantity of garden soil, sand and compost or leaf mould. Alternatively, 30 per cent soil, 20 per cent compost, 20 per cent coco-peat, 20 per cent sand and 10 per cent neem cake powder can be mixed to get ideal pot mixture.

During summer months, irrigation is provided at weekly interval to keep the soil moist. However, care should be taken to provide proper drainage during *kharif* season. Hand weeding can be done at 30 days after planting to keep the plot weed free. The weeds are to be removed from the pot/container for better growth of the plant.

Shoot and fruit borer is the most harmful insect of brinjal plant causing significant yield loss. The symptoms of damage are withering of terminal shoots, holes on shoots & fruits, dropping of flower buds and drying of leaves. It can be controlled by removing the affected shoots and fruits, installation of pheromone trap, spraying of neem seed kernel extract (NSKE) and application of neem based pesticides. Besides, insecticides like dimethoate or emamectin benzoate or flubendiamide can be applied to control the insect. Stem borer is another harmful insect of brinjal plant. The prominent symptoms are drooping of top shoots, stunted growth and poor fruit setting. The control measures include destroying the affected plant parts, use of light trap, spraying of neem oil and application of prophenofos or dimethoate.

Brinjal crop is affected by various diseases. Bacterial wilt is a common disease in brinjal. Stunted growth of plant, yellowing of leaves and sudden wilting of the plant are the characteristic symptoms of this disease. Use of resistant/ tolerant varieties, judicious use of water and spraying of plantomycin along with copper oxychloride can control this disease. Tobacco mosaic virus (TMV) is another disease affecting brinjal crop. The typical symptoms are mottling of leaves, stunted growth of plants and deformation of leaves. The disease is transmitted by insect vectors. To control this disease, the plots should be kept clean and the infected plants are to be removed & destroyed. The insect vectors can be controlled by spraying of insecticides like dimethoate @ 2 ml/ litre or metasystox @ 2 ml/litre of water. Brinjal is also affected by the fungal disease alternaria leaf spot. In this case, spots with concentric rings appear on the leaf surface. The spots are irregular with 4-8 mm in diameter at initial stage and spread to large areas of the leaf at later stage. Severely affected leaves may drop off. Spraying of copper oxychloride or carbendazim can effectively control the disease.

Brinjal is harvested at about 55-60 days after transplanting. Harvesting is done at 4-5 days interval. The yield of the crop varies depending on variety, management practices and local climatic condition. The high yielding varieties can yield up to 25-30 kg of fruits from an area of 10 m^2. Generally, a brinjal plant can produce about 15-20 fruits in a season. Of course, it varies depending on variety and management practices.

Capsicum

Capsicum is also known as sweet pepper or bell pepper. It is a popular vegetable grown during winter season. It can be grown round the year in higher altitude having low temperature. The ideal daytime temperature is 26-28^0 C and night temperature is 16-18^0 C for growth and development of capsicum. Very low night temperature of less than 12^0 C is not favourable for the plant. Capsicum cannot withstand heavy rain during flowering stage. In home gardens, capsicum can be grown in small patches in the backyard areas. In absence of surplus land, it can be planted in pots, containers or polythene bags placed on the rooftop or terrace. About 10-12 plants are adequate to fulfil the requirement of a small family. There are several high yielding varieties and hybrids grown in different parts of the country. Some of the high yielding varieties are Arka Basant, Arka Gaurav, Arka Mohini, etc. The promising hybrids are Arka Athulya, Indra Capsicum, Green Gold, Bharath, Sun 1090, etc.

Capsicum gives a good yield when grown in well drained sandy loam soil with adequate organic matter. Soil with poor drainage facility and high acidic reaction is not suitable for this crop. The ideal soil pH ranges between 5.5 and 6.8 for cultivation of capsicum. Capsicum can be planted during the months of September to November. The seed requirement is 0.5 gram for 10 m^2 area. The seeds are to be treated with carbendazim or *Trichoderma viride*or *Pseudomonas fluorescens* to prevent infestation of various diseases. The seeds are sown in raised beds to grow seedlings. Sowing is done in lines at a spacing of 2.5 cm followed by covering with farm yard manure or soil. Watering is done daily with rose can. Seedlings can also be raised in pro tray for higher percentage of germination. The pro trays are filled with a mixture prepared with sterilized coco peat and neem cake along with *Azospirillum* and phosphobacteria. The seedlings are ready for transplanting at 3-4 weeks after sowing.

The field is prepared to make the soil a fine tilth. It can be done by spade or other implements. Application of FYM or compost may be done @ 25 kg in 10 m^2 area. About 250 gram of neem or karanja oilcake can be applied in an area of 10 m^2 area. Ridges and furrows are made at distance of 45-60 cm. Seedlings are planted on the ridges at plant to plant spacing of 30 cm. Gap filling, if required, can be done 7 days after transplanting. Fertilizers are to be applied based on soil test values. The general recommendation is 120 gram N, 80 gram P_2O_5 and 50 gram K_2O for capsicum grown in 10 m^2 area. All the P_2O_5 and K_2O along with one third of nitrogen is to be applied as basal dose. Rest of N is top dressed in equal quantity at 30 and 60 days after transplanting. Depending on the soil moisture condition, irrigation may be done at 7-10 days interval. Weeds can be controlled manually by uprooting or by hoeing the field at three weeks after transplanting. In case of pot culture, the space around the plant should be kept clean for better plant growth.

Capsicum is affected by various diseases like bacterial wilting and powdery mildew. Damping off disease is observed in the nursery. There is sudden death of plants due to bacterial wilt. This can be controlled by spraying of plantomycin along with copper oxychloride or carbendazim. In case of powdery mildew disease, there is appearance of white substances on the leaves. This can be controlled by spraying of fungicides like propiconazole, carbendazim, etc. Damping off disease causes death of seedlings in the nursery. This can be controlled by soil drenching of copper oxychloride or mancozeb. Fusarium wilt is another fungal disease affecting this crop. The symptoms are yellowing of foliage and wilting of upper leaves. This can be controlled by application of fungicides like propiconazole or carbendazim in the crop. Capsicum is affected by several insects like aphids, thrips, etc. The insects suck the cell sap and the plants get stunted. This can be controlled by application of profenophos or imidachloprid. Application of neem/karanja oilcake or spraying of neem based pesticides can prevent the spread of insect or diseases.

Capsicum is harvested before ripening at green fruit stage. From an area of 10 m^2, average yield of 15 kg can be obtained. However, more yield can be obtained from hybrids.

Chili

Chili is an essential ingredient of Indian cuisines. It is one of the most valuable crops of India being used as spice, condiments, sauce and pickle. Dried chilies are used for preparation of spice pastes and curry powders. Pungency in chili is caused by the active constituent capsaicin. Chili can be grown round the year in fields, kitchen gardens or in pots. It grows well in warm and humid climate with a temperature range of 20-30°C. Two to three plants in the backyard or in the containers placed on the rooftop or terrace can fulfil the green chili requirement of a family. The predominant high yielding chili varieties cultivated across the country are Pusa Sadabahar, Pusa Jwala, Pant C-1, Pant C-2, Sindhur, Ujjwala, Utkal Ava, Jawahar 218, Arka Lohit, Co 1, etc. Some of the hybrids are Tejaswini, Agniban, Champion, Surya, etc.

Chili can be grown in a variety of soil ranging from sandy loam to clayey soil. But, well drained loamy soils with good water holding capacity is best suited for this crop. The ideal pH range lies within 6.0-7.0.

Chili requires thorough land preparation to obtain a fine soil tilth. During final land preparation, 15-20 kg KYM or compost and 250 gram neem or karanja oilcake may be mixed with the soil in an area of 10 m^2. Application of neem/karanja oilcake can protect the crop from white ants and other soil borne pest. Application of *Azospirillum* and Phosphobacteria along with FYM helps in better plant growth. The crop can be planted during June-July as *kharif* crop, January-February as summer crop and October - November as *rabi* crop.

Chili is sown after raising of seedlings in nursery bed. The nursery bed is provided with compost or farm yard manure for better growth of seedlings. The seeds are to be treated with *Trichoderma* or *Psuedomonas* to prevent seed borne diseases. The seeds are sown in nursery bed with a row spacing of 10 cm. The seeds germinate in 5 to 7 days. The nursery bed is drenched with copper oxychloride @ 2.5 g/l of water at 15 days' interval to prevent occurrence of damping off disease. The nursery bed is to be watered daily using rose can. The seedlings are ready for transplanting at 21-30 days after germination. Seedlings can also be raised in pro-trays @ 1 seed per cell. Good quality seedlings can also be procured from nearby nurseries for transplanting. The seedlings are transplanted in the bed with a spacing of 60 cm X 45 cm for high yielding varieties and 75 cm X 60 cm for hybrids.

Chili can also be grown in pots for domestic consumption. Pot mixture can be prepared by taking equal quantity of coco peat, compost, sand and garden soil. The potting mixture is to be soaked overnight in water in a pot. The chili seeds are placed in the pot and watering is done daily till germination of seeds. After appearance of leaves, the seedlings are to be transferred to bigger pot of 15 cm.

Fertilizer dose should be decided based on soil test values. In general, a nutrient dose of 120 gram nitrogen, 60 gram phosphorous and 60 gram potassium is applied per 10 m^2 area. Out of the total nitrogen, 30 gram is applied as basal along with entire quantity of phosphorous and half of potassium. At 30 days after transplanting, 60 gram nitrogen is to be top dressed. Rest of nitrogen and potassium are to be top dressed at 60 days after transplanting. Application of potassium in the form of potassium sulphate will increase quality of chili. Application of micronutrients improves the yield and quality of chilli. Foliar spray of Zn SO_4(0.5 %) thrice at 10 days interval from 40 days after planting and spraying of Mn (1 %) at 60 days after planting are beneficial. Spraying of growth regulator like NAA @10 mg/l of water at 60 and 90 days after planting increases fruit yield.

The plot or field should be maintained weed free by hand weeding or hoeing. Adequate soil moisture should be maintained for a successful crop growth. The crop should be irrigated in weekly intervals. Low moisture in soil during flowering and fruit formation results in flower and fruit drop. Excessive rainfall or standing water is harmful to the crop as it causes rotting of the plant root.

Chili plant is affected by various insects and diseases. The major insects are thrips, aphids, white fly, etc. The plant growth is severely affected due to attack of thrips and aphids as these insects suck the cell sap from the plant. This can be controlled by application of imidachloprid or dimethoate. The plant is affected by leaf curl disease and powdery mildew disease. The symptoms of leaf curl disease are inward curling of leaves, stunted plant growth and shortened internode. The virus causing leaf curl disease is generally transmitted by whitefly. The affected plants need to be removed and thrown away. Controlling the insect by application of profenophos or fenvalarate can minimize the spread of the disease.

The symptoms of powdery mildew disease are appearance of white powdery growth on lower side of the leaves and shedding of foliage. This can be managed by spraying of fungicides like propicocazol, mancozeb or copper oxychloride. Spraying of wettable sulphur 0.25% can also control this disease.

Harvesting can be done when the fruits are fully grown and tightly filled. The fruits are plucked at red ripe stage for use as dry chili. In some cases, the initial pickings are done for green chili and produce from subsequent pickings are used as dry chili. The yield of green chili is about 5-6 kg from an area of 10 m^2. The yield can be substantially higher by growing hybrids with proper management practices.

Potato

Potato is the most important vegetable consumed across the world. It is grown during winter season in most of the states of India. However, it can also be grown during *kharif* season in hilly tracts. Potato provides a source of low cost energy in human diet. Potato is a rich source of starch, vitamins and minerals. It contains about 20.6 % carbohydrate, 2.1 % protein, 0.3 % fat, 1.1 % crude fibre and 0.9 % ash. It is grown under moderately cool climatic condition. The ideal temperature favouring vegetative growth is around 24°C, while tuber development is better at about 20° C temperature. The crop can be raised in diverse climatic condition varying from plains to an altitude of 3000 m above the sea level. In home gardens, it can be grown in small plots to produce potato for home consumption.

The popular varieties of potato grown across the country include Kufri Jawahar, Kufri Chandramukhi, Kufri Sutlej, Kufri Bahar, Kufri Anand, Kufri Ashoka, Kufri Pukhraj, Kufri Jyoti, Kufri Megha, Kufri Lauvkar, Kufri Swarna, Kufri Thangam, Kufri Malar, Kufri Soga, Kufri Giriraj, etc.

Potato performs well in sandy loam soil, rich in organic matter with good drainage facility. The optimum soil pH ranging from 5.5 to 7.5 is suitable for this crop. It cannot be taken in saline and alkaline soils. Usually potato is planted during October-November as *rabi* crop in plains. In hilly areas, it can also be planted during June-July as kharif season crop.

Potato crop is commonly raised by planting of tubers. The tubers used for seed purpose should be of uniform size and free from disease. Sprouting of tubers are to be ensured before planting. It is better to use the entire seed tuber for planting. However, large size tubers can be cut in to pieces weighing 20-30 g with at least one active eye bud. The seed tubers are to be treated with fungicide before planting. The recommended seed rate is about 1.5-2.0 kg for an area of 10 m^2.

Before planting, the land should be properly tilled to make the soil fine tilth. Application of 15 kg FYM or compost and 250 gram of neem or karanja oilcake is to be done for an area of 10 m^2 at the time of soil preparation. Furrows are to be made at a spacing of 45 cm either by hand hoe or ridger. The seed tubers are to be planted at a distance of 20 cm.

Azospirillum and Phosphobacterium may be mixed with FYM and applied at the time of planting. Fertilizer dose of 120-150 gram N, 60-80 gram P_2O_5 and 120-150 gram K_2O is required for 10 m^2 area. All phosphorous and half of nitrogen & potassium are applied as basal fertilizer at the time of planting. Rest of the nitrogen and potassium are to be top dressed at the time of earthing up at about 25-30 days after planting. However, fertilizer may be applied basing on soil test value. If required, boron and zinc may be provided for better tuberization. The crop is to be irrigated just after planting. Subsequent irrigations should be given in 7-10 days interval depending on soil moisture status. While irrigating the field, care should be taken to fill up about two third of the furrows. Deficit of water at germination, tuber formation stage and tuber development stage adversely affects the yield.

Weeds compete with the crop for nutrition and water. Hence, timely steps are to be taken to control the weeds. The critical period of weed-competition is up to 45 days after planting. Weeding can be done by hoeing the field without disturbing the stolons. If required, another hoeing is to be done followed by earthing up.

Potato is affected by various diseases and pests. Some of the diseases are common scab, early blight and late blight. Common scab of potato is a soil-borne disease. The symptoms of common scab are scab-like raised or slightly sunken rough and corky patches on tuber. This can be prevented by using seed tubers treated with 3% boric acid for 30 minutes. This disease is more prevalent in alkaline soil. In case of blight, pale green irregular water soaked spots appear on the leaves. There is formation of white mildew ring on the lower side of the leaves. This can be controlled by spraying Mancozeb (2 g/lit) or Chlorothalonil (2 g/lit) or Metalaxyl (2.5 g/lit). Growing resistant varieties like Kufri Sutlej, Kufri Badsah and Kufri Jawahar can prevent spread of this disease.

Insects like cutworm, aphid and white grub affect the crop. Aphids can be controlled by spraying Methyl demeton 25 EC or Dimethoate 30 EC @ 2 ml/lit. Drenching of the collar region of the plants with profenophos @ 2 ml/lit can control cutworm. Drenching of soil with insecticide can control white grub. Insect and disease can be controlled by spraying of neem oil or neem based pesticides. Installation of light trap or yellow sticky trap can also control the insects.

The time of harvest is very important for potato. The crop is ready for harvest when the plants are completely died. The average yield of 25-30 kg can be obtained from 10 m^2 area. Higher yield can be obtained with proper management practices.

Pumpkin

Pumpkin is an important cucurbitaceous crop grown during kharif and summer season. It is a highly productive crop with good nutritive value and longer storability. It is rich in vitamins, minerals and antioxidants. The fruit, young leaves, tender stems and flowers are used as vegetables. It can be grown in diverse climatic condition and requires bright sunshine. The ideal temperature for pumpkin cultivation is 18-30^0 C. In home gardens, few pits can be planted to fulfil the requirement of family. It can also be planted in pots, container or polybags and kept on the rooftop or terrace. There are good numbers of high yielding varieties grown across the country. Some of those are Co 1, Co 2, Pusa Vikash, Baidyavati, Arka suryamuki and Arka Chandan. The promising hybrids are Pusa Hybrid 1, MPH 1, etc.

Pumpkin can be grown in various types of soils. It does well in sandy loam to loamy soil with adequate organic matter and good drainage. The pH range of 5.5-7.0 is ideal for this crop. The *kharif* crop is sown during June-July and summer crop is sown during January-February. However, in some areas it can also be taken up as a *rabi* crop.

The seeds are to be treated with carbendazim or thiram before planting. Alternatively, it can also be treated with *Trichoderma viridae* to prevent seed borne diseases. In cooler areas, the seeds can be soaked in water for quicker germination. Pumpkin is usually planted in the pits of 30 cm x 30 cm x 30 cm size dug at a spacing of 1.5 m x 1.5 m. Three to four seeds are planted in a pit. In some areas, pumpkin is planted on ridges with a plant to plant spacing of 60-75 cm.

Application of 5-6 kg of FYM along with 8 gram nitrogen, 8 gram phosphorus and 8 gram potassium is to be applied per pit at the time of planting. Subsequently, 8 gram nitrogen and 8 gram potassium is applied in each pit after 30 days of planting. *Azospirillum* and *Phosphobacteria* may be applied to enhance plant growth. Application of neem cake in the pits will protect the crop from termite attack. Application of zinc and boron will improve the plant growth and increase the yield. Spraying of etherel will help in inducing more number of female flowers. Weeds are to be removed at weekly intervals commencing from 10 to 15 days after planting. The crop is irrigated at an interval of 6-8 days depending on soil moisture content. The critical periods of water requirement are flowering and fruit development stage. Water or nutrient deficit may increase the number of male flowers, thereby resulting in less number of fruits per plant.

Pumpkin is generally affected by powdery mildew, downy mildew and yellow mosaic virus diseases. White to grey spots appear on the leaves in case of powdery mildew infestation. There may be deposition of white powdery mass on the fruits, when the infestation is severe. In case of downy mildew disease, angular yellow spots are seen on the lower surface of the leaf and eventually the affected leaves die. Powdery mildew and downy mildew diseases can be controlled by spraying of mancozeb or chlorothalonil @ 2 g/litre water twice at 10 days' interval. When the plant is affected by yellow mosaic virus, the interveinal space of the leaves become yellow. This is a viral disease being spread by the insect vector. It can be managed by spraying insecticides like malathion or profenophos to control the insect.

Red pumpkin beetle, fruit fly and thrips are common insects affecting pumpkin. The symptoms of red pumpkin beetle are damage of leaves and plant part touching the ground. It can be controlled by spraying carbaryl (0.2%) of quinalphos (0.05%). The young fruits start rotting due to attack of fruit fly. Growing of resistant varieties such as Arka Suryamukhi is advisable to prevent insect attack. Foliar spray of cartap hydrochloride @ 1.5g/lit of water can control this insect. Thrips suck the cell sap and the plants get stunted. Thrips can be controlled by spraying of imdachloprid @ 2.5 ml/10 lit of water. Spraying of neem or neem based insecticides can also control attack of disease and insect.

Harvesting can be done when the fruits turn yellow and easily detached from the stem. Well matured fruits can be stored for longer period. Average yield of 12 to 15 kg of fruits can be obtained from each pit. The yield can be more with good management practices.

Bottle Gourd

Bottle gourd is a popular vegetable, largely used for culinary purpose. Its medicinal properties have made it popular among Indian families. The hard cells are used as utensils and preparation of musical instruments. It is low in saturated fat & cholesterol and high in dietary fibre, vitamin C, vitamin K, riboflavin, zinc, thiamine, iron, magnesium & manganese. The ideal time for growing bottle gourd is kharif and summer seasons. Bottle gourd performs well under moist and hot climatic conditions. The suitable temperature range for its cultivation is 20-32°C. The day temperature above 40° C may result in scorching of leaves. Similarly, temperature below 10°C adversely affects germination and plant growth. This vegetable cannot withstand frost condition. It can be taken up in the backyard or any available space in the home gardens. The most common varieties grown across the country are CO 1, Punjab Komal, Phule BTG 1, Pusa Summer Prolific Long, Pusa Summer Prolific Round, Arka Bahar, Pusa Manjari, Pusa Meghdoot, etc. Some of the promising hybrids are PBOG 1, NBGH 22, NDBG 4, etc.

Bottle gourd can be grown in all types of soil. However, it grows well in sandy loam soil with good fertility status. The soil pH ranging from 6.5 to 7.5 is suitable for this crop. The soil should have good drainage during rainy season. Soils with highly acidic, alkaline or saline reactions are not suitable for bottle gourd cultivation. The land is prepared to get a fine soil tilth for better germination and growth of the crop. Bottle gourd is planted in pits of 30 cm x 30 cm x 30 cm size spaced at a distance of 1.2 meter either side. Prior to sowing, the seeds are treated with *Trichoderma viride* or *Pseudomonas fluorescens* or Carbendazim to prevent seed borne diseases. About 3-4 seeds are planted in each pit and the seedlings are thinned after 15 days of germination to keep two seedlings per pit. In case of hybrids, two seeds can be planted per pit.

Basal application of 4-5 kg farm yard manure or compost per pit will help better initial growth of the plant. Fertilizer is to be applied as per the soil test results. However, about 7 gram nitrogen, 4 gram phosphorus and 7 gram potassium are required for each pit. All the phosphorus & potassium and half of nitrogen are to be applied at the time of planting. Rest of the nitrogen is to be top-dressed at three weeks after planting. *Azospirillum* and Phosphobacteria can be mixed with FYM and applied before planting to get more number of good quality fruits. Besides, application of neem oil cake in the pits will prevent termite attack. Foliar spray of boron will accelerate plant growth and enhance fruit production. The plants are provided with support through trellis made up of bamboo stick or dried branches for better growth and fruit setting. It also keeps the fruits away from soil surface, which otherwise would lead to fruit rot. One pre-planting irrigation should be provided to ensure adequate soil moisture for better germination. Subsequently, water is supplied at weekly intervals in summer season. During kharif season, water is supplied depending on soil moisture condition and adequate care is to be taken to drain out excess water from the plots. Moisture stress during top dressing, flowering and fruit setting may lead to drastic yield loss. Weeding can be done manually as and when necessary.

Bottle gourd is affected by powdery mildew and downy mildew diseases. In case of powdery mildew disease, white spots appear on the leaves. In case of downy mildew infestation, angular yellow spots are seen on the lower surface of the leaves. Powdery mildew and downy mildew diseases can be controlled by spraying of mancozeb or chlorothalonil @ 2 g/litre twice at 10 days' interval.

Beetles, fruit flies and caterpillars are the common insects affecting bottle gourd. The beetles feed on root and plant parts touching the ground. The young fruits start rotting due to attack of fruit fly. Long shining caterpillars feed on the leaves. These insects can be controlled by spraying of malathion 50 EC or dimethoate 30 EC or methyl demeton 25 EC @ 2 ml/lit. Application of neem oil or neem based pesticide can also control the insects and diseases.

Bottle gourd is harvested when the fruits are at tender stage. The market value of this crop reduces if harvest is delayed. Bottle gourd can yield about 5-6 kg fruits from every pit. Providing stake to the plant can further enhance the yield.

Bitter Gourd

Bitter gourd is an important vegetable grown across the country. As the name indicates, the taste of the fruit is bitter and is widely used to control of blood sugar. Hundred grams of raw bitter gourd contain about 0.15 gram fat, 0.93 gram protein and 2.6 gram fibre. It is also an important source of vitamin A, vitamin C, iron, zinc and potassium. It performs well at temperature range between 30-35^{0}C. It cannot be grown at temperature below 18^{0}C or above 40^{0}C. Bitter gourd grows well in warm and humid climate. The important varieties under cultivation are Arka Harit, Pusa Do Mausmi, Priya, Preethi, Pusa Vishesh, Coimbatore Long, etc. In home gardens, it can be planted in the available space or near the boundary for home consumption. It can also be grown on the rooftop or terrace with the help of pot, container or polybags. Staking can be provided for trailing the plant to obtain higher yield.

Like other vegetables, bitter gourd grows in well drained sandy loam soil with adequate organic matter. The ideal pH requirement is 6.5-7.5. It is cultivated during *kharif* and summer seasons. The *kharif* crop is sown during June-July and summer crop is sown during January-February. The seeds are to be treated with carbendazim or thiram to prevent seed borne diseases. Seeds can also be treated organically by using *Trichoderma viride* or *Pseudomonas fluorescens*.

Soil should be properly tilled to obtain fine tilth. Pits of 45 cm x 45 cm x 45 cm size are prepared at a spacing of 1.2 m x 1.2 m. About 3-4 seeds are to be sown in each pit. Normally, seeds are sown to grow bitter gourd. Sometimes, seedlings are raised in polythene bag to ensure proper plant stand. Polybags of 10 cm diameter and 10 cm height are used for sowing the seeds. About 15 days old seedlings are transplanted in the main field. Basal application of 4-5 kg farm yard manure or compost per pit will help better initial growth of the plant. Fertilizer is to be applied as per the soil test results. However, about 8 gram nitrogen, 4 gram phosphorus and 4 gram potassium is required for each pit. All the phosphorus & potassium and half of nitrogen are to be applied at the time of planting. Rest of the nitrogen is to be top-dressed at three weeks after planting. *Azospirillum* and Phosphobacteria can be mixed with FYM and applied before planting to get more number of good quality fruits. Besides, application of neem or karanja oilcake in the pits can prevent termite attack. Foliar spray of boron accelerates plant growth and enhances fruit production.

Irrigation is given to the pits immediately after sowing for better germination. Subsequently, irrigation is given depending on soil moisture condition. In summer season, irrigation is required at 7-10 days' interval. Adequate drainage facility is to be provided during *kharif* season to protect the crop from heavy rain or water stagnation. Trailing the vines on bamboo or wooden sticks facilitates better plant growth, reduces disease infection and enhances yield. Now-a-days, plastic wires or metallic wires are used to erect trellis for providing support to the plants.

Weeds can be controlled by uprooting or hoeing the soil about three weeks after planting. Hoeing also helps in better root growth. Spraying of 100 ppm ethrel can result in better fruiting.

Bitter gourd is affected by various insects like aphids, beetles and caterpillars. The aphids suck the sap from the plant resulting in stunted growth of plant. The beetles and caterpillars feed on the leaves and young twigs causing damage to the plant. Aphids can be controlled by spraying of Imidachloprid or profenophos. Beetles and caterpillar can be controlled by spraying dimethoate or methyl demeton @ 2ml/lit.

This crop is affected by various fungal diseases like powdery mildew, downy mildew and leaf spot. There is deposition of white substances on the leaves due to infestation of powdery mildew and downy mildew. Brown spots are observed in case of leaf spot disease. These can be controlled by spraying mancozeb or chlorothalonil 2 g/lit twice at 10 days interval.

Bitter gourd is harvested at tender stage. Usually, harvesting is started at 60-70 days after planting. About 2 kg fruits can be obtained from each plant. Higher yield can be obtained with hybrids grown under good management practices.

Cucumber

Cucumber is a popular vegetable mostly consumed as raw, salad or in cooked form. Cucumber contains 96% water and is refreshing when consumed during summer season. It is a good source of potassium, fibre, vitamins minerals, etc. It has multifarious health benefits and is used as a hydrating food. This can be grown in the kitchen garden, rooftop or in the terrace. Available area in the backyard can be utilized for planting cucumber. The optimum temperature for growing cucumber is 25-35°C. This crop requires a moderate warm temperature. It is normally grown during *kharif* and summer seasons. The *kharif* crop is sown during June-July and summer crop is sown during January-February. The promising cucumber varieties are Japanese Long Green, Straight Eight, Pusa Sanyog, Poona Khira, Supriya, Poinsettee, Sheetal, etc.

Cucumber can be sown in wide range of soils starting from sandy loam to heavy soil. But loamy soil with high organic matter and good drainage is best for the crop. The ideal pH range for cucumber is 5.5 -6.8. Prior to sowing, seeds are treated with *Trichoderma viride* or *Pseudomonas fluorescens* or Carbendazim to protect the crop from diseases.

Cucumber requires well pulverised weed free field. To get fine soil tilth, land should be thoroughly tilled before planting. FYM or well decomposed compost may be applied @ 3-4 kg per pit. Pits should be prepared with a size of 30 cm X 30 cm X 30 cm at a spacing of 1.2 m X 1.2 m. Neem or karanja cake should be applied to minimize attack of termite and other insects. Three to four seeds are to be sown in each pit. Fertilizer should be applied based on soil test results. In absence of soil test results, about 12 gram nitrogen, 6 gram phosphorus and 6 gram potassium can be applied in each pit. All the phosphorous and potash along with half of nitrogen is to be applied at the time of planting. The balance quantity of nitrogen is to be top-dressed at 25-30 days after sowing. Application of *Azospirillum* and Phosphobacteria before sowing helps to get better quality produce. Spraying of boron and zinc helps in better plant growth and fruit yield. The pits should be made weed free by hoeing or manually uprooting the weeds. Earthing up helps in better root growth and facilitates water management.

The critical periods of water requirement are germination, flowering and fruit setting. Irrigation is given at 7-10 days' interval as per soil moisture status during summer season. As cucumber cannot withstand water stagnation, care should be taken to drain out excess water from the field during rainy season.

Cucumber is affected by various insects like fruit fly, aphids and caterpillars. The fruit flies enter the fruit at early stage and damage the fruit. The aphids suck the sap from the plant resulting in stunted growth of plant. The caterpillars feed on the leaves and tender parts causing damage to the plant. Fruit flies can be controlled by foliar spray of 3.0 % *neem* oil or neem based pesticide. Aphids can be controlled by spraying Imidachloprid @ 0.25 ml or Dimethoate @ 2 ml/litre water. Caterpillar can be controlled by spraying of profenophos or Methyl demeton @ 2ml/litre water. Downy mildew is the most common disease of cucumber. This can be controlled by spraying mancozeb or chlorothalonil @ 2 g/litre water twice at 10 days' interval.

Depending on local climatic condition and variety, harvesting starts at 45-60 days after sowing the seed. On an average, 8-10 harvests can be done during harvest period. The quantity of production varies depending on the variety, growing season and management situations. However, about two kilogram fruits can be obtained from one pit. Similarly, from a smaller area of $10m^2$, 8 - 10 kg fruit can be obtained.

Ridge Gourd

Ridge gourd is a popular vegetable grown throughout the year. It is a fleshy vegetable used in preparation of curry. It is highly nutritious and contains vitamins, minerals and dietary fibre. Ridge gourd is having medicinal properties and beneficial for control of blood sugar & digestive problem. Ridge gourd thrives well in hot and humid climate. The ideal temperature for ridge gourd cultivation is 25° to 35°C. The most common varieties of ridge gourd are Pusa Nasdar, Pusa Nutan, Pusa Chikni, Pant Torai, CO 1, PKM 1, Swarna Manjari, Punjab Sadabahar, Satputia, Arka Sujat, Desi Chaitali, Phule Sucheta, Kankan Harita, etc.

Ridge gourd can be grown in wide variety of soils. However, it grows well in sandy loamy soil. Suitable soil pH for ridge gourd cultivation ranges from 6.5 to 7.5. The soil should be prepared thoroughly to get good soil tilth. This crop requires good drainage condition and cannot tolerate standing water. Better yield and quality of produce can be obtained by applying organic matter or farm yard manure @ 3-4 kg per pit. Seeds should be treated with *Pseudomonas fluorescens*or *Trichoderma viridae*. Sowing of ridge gourd is done during June-July for kharif season crop and January-February for summer crop.

The seeds are sown by dibbling in pits made at a spacing of 1.2 m X 1.2 m. Three to four seeds are sown in each pit. The seeds can also be sown in polybags or pots in backyard garden or rooftop. During the early stage of plant growth, shallow hoeing is preferred. The plants should be provided with a suitable support by bamboo or wooden sticks to facilitate trailing.

Fertilizer should be applied based on soil test results. Alternatively, about 12 gram nitrogen, 6 gram phosphorus and 6 gram potassium can be applied in each pit. All the phosphorous and potash along with half of nitrogen is to be applied as basal fertilizer at the time of planting. The balance quantity of nitrogen is to be top-dressed at 25-30 days after sowing. Application of *Azospirillum* and Phosphobacteria can be done by mixing with FYM to get better quality produce. Spraying of boron and zinc helps in better plant growth and higher fruit yield. Weeds can be controlled manually by uprooting or hoeing the pits. Earthing up after three weeks of planting will facilitate irrigation and drainage. The critical period of water requirement is germination, flowering and fruit setting. One pre-sowing irrigation can be given prior to sowing of seeds for better germination. Subsequently, irrigation may be provided at weekly interval. However, the summer crop requires frequent watering at an interval of 5 to 6 days depending on soil moisture condition. The rainy season crop generally does not require any irrigation. But, care is to be taken to avoid water stagnation.

Ridge gourd is affected by various insects like fruit fly, aphids and caterpillars. The fruit flies enter the fruit at early stage and damage the fruit. The aphids suck the sap from the plant resulting in stunted growth. The caterpillars feed on the leaves and tender plant parts causing damage to the plant. Fruit flies can be controlled by using neem oil or neem based pesticide as foliar spray. Aphids can be controlled by spraying of Imidachloprid @ 0.25 ml/litre water. Caterpillar can be controlled by spraying profenophos or Dimethoate @ 2 ml/ litre water. Downy mildew is the most common disease of ridge gourd. There is deposition of white substances on the leaves due to infestation of this disease. This disease can be controlled by spraying of Mancozeb or Chlorothalonil 2 g/ lit twice at 10 days' interval.

Harvesting of ridge gourd starts at 45 to 60 days after sowing. The full grown fruits are harvested by cutting with knife or plucking. Depending upon the variety and management condition, about 2 kg fruit can be obtained from a plant.

Pointed Gourd

Pointed gourd or parwal is a popular vegetable grown in various parts of India. It gives good income to the farmers due to comparatively higher market price of the produce. Pointed gourd is a healthy vegetable and good source of vitamin. It does well in warm to hot humid climate. It cannot tolerate frost or extreme cold condition. The plant requires 30-35 °C temperature for optimum growth and development. Some of the popular varieties of pointed gourd are Swarna Rekha, Rajendra Parwal-1, Rajendra Parwal-2, Swarna Alaukik, etc.

Well-drained sandy loam to clay soil is suitable for cultivation of pointed gourd. The optimum soil pH range of 6.0 to 6.5 is best for pointed gourd cultivation. This crop does not grow well in heavy soils, but performs well in alluvial light soil. Pointed gourd can be grown in home gardens either on the ground or on the rooftop or terrace. Growing the plant with support of staking or trellis can enhance the yield. Normally, pointed gourd is propagated through vine cuttings and root cuttings. Pointed gourd is a dioecious plant with separate male and female plants. While planting, ratio between female and male plant should be 9:1. As a thumb rule, farmers identify the sex of the plant by the texture of the root surface. The roots of female plants are smooth whereas, those of male plants are rough. However, this may vary depending on variety and soil condition. The female: male ratio can be maintained precisely after emergence of flowers. Root/vine cuttings are to be treated with fungicide before planting to prevent soil borne diseases.

Pointed gourd can be cultivated throughout the year. However, October-January is the best time for planting of pointed gourd. In some areas, planting can be extended up to February-March. Land should be prepared thoroughly until the soil reaches a fine tilth. Application of 4-5 kg FYM or compost is to be done per pit to enhance soil fertility. The cuttings are to be planted in pits made at a spacing of 1.2 m X 1.2 m.

Fertilizer should be applied based on soil test value. However, about 18 gram nitrogen, 12 gram phosphorus and 12 gram potassium can be applied in each pit. All the phosphorous and potash along with half of nitrogen is to be applied at the time of planting. The balance quantity of nitrogen is to be top-dressed in two equal splits at one month and two months after planting. Application of *Azospirillum* and Phosphobacteria can be done by mixing with FYM to get better quality of produce. Spraying of boron and zinc helps in better plant growth and fruit yield.

Weeds may be controlled manually by hoeing. Earthing up is necessary at three weeks after sowing for better plant growth. The critical periods of water requirement are flowering and fruit setting. One irrigation is given just after planting of the root cuttings. Subsequently, irrigation may be given at 7-10 days interval depending on soil moisture content. However, the summer crop requires frequent watering at regular intervals. During rainy season, adequate care is to be taken for draining out excess water.

Pointed gourd is affected by various insects. The aphids suck the sap from the plant resulting in stunted growth. The caterpillars feed on the leaves and tender plant parts causing damage to the plant. These insects can be controlled by foliar spray of neemoil or neem based pesticide. Chemically, the insects can be controlled by spraying of Imidachloprid or profenophos. Insects can also be controlled by use of yellow sticky trap or light traps. Pheromone trap can also be used to control insects. Downy mildew is the most common disease of pointed gourd. There is deposit of white substances on the leaves due to infestation of downy mildew. This can be controlled by spraying propiconazole or tebuconazole.

Pointed gourd is harvested before complete maturity. The full grown fruits are harvested by plucking. Depending upon the variety and management condition, about two kg fruit can be obtained from each pit.

Watermelon

Watermelon is an important cucurbitaceous crop grown in various parts of India. It is an excellent dessert fruit and contains about 90 per cent water along with protein, minerals and carbohydrates. Watermelon is rich in vitamins and other antioxidant compounds, which contribute immensely to human health. Being a warm season crop, the plant requires ample of sunshine and dry weather for growth and fruiting. This crop is sensitive to frost damage. It grows well within temperature range of 20-25°C. The soil temperature should be in the range from 18 to 20°C for better root development. But, extreme cold is harmful for this crop. A few plants can be grown in the house premises or in the kitchen garden to get good quality watermelon. Several high yielding varieties and hybrids of water melon are grown by the farmers. Some of the high yielding varieties are Sugar Baby, Asahi Yamato, Black Magic, Arka Akash, Arka Muthu, Arka Manik, Durgapur Kesar, etc. Some of the hybrids are Arka Jyoti, Pusa Bedana, Madhuri 64, Varun, Millionaire, Sweet Heart, Madhu, Milan, etc.

Water melon performs well in fertile sandy loam soil having good drainage facility. The ideal pH range for growing water melon is 6.5-7.5. River bed with alluvial soil is the most suitable site for cultivation of watermelon. Before sowing, seeds are to be treated with *Trichoderma viridae* or *Pseudomonas fluorescens*. Watermelon is normally sown during the months of November-December. In north India, sowing can also be done during February-March.

The crop requires fine soil tilth for better growth and production. Well decomposed FYM or compost is applied @ 4-5 kg per pit during land preparation. Pits of size 30 cm X 30 cm X 30 cm are prepared with a spacing of 1.2 m X 1.2 m for sowing the seeds. Application of neem or karanja oilcake can minimize termite attack. Three to four seeds are to be sown per pit to ensure adequate plant stand.

Fertilizer should be applied based on soil test results. However, about 9 gram nitrogen, 6 gram phosphorus and 14 gram potassium are required for each pit. All the phosphorous and potash along with half of nitrogen are to be applied at the time of sowing. The balance quantity of nitrogen is to be top dressed at 20-25 days after sowing of seeds. It is advised to complete fertilizer application before fruit setting. Application of *Azospirillum* and Phosphobacteria being mixed with FYM before sowing is helpful to get quality fruits. Spraying of boron and zinc helps in better plant growth and fruit yield. Weeds can be controlled by hoeing the pits at three weeks after sowing. The critical periods of water requirement are germination, flowering and fruit setting. Irrigation is given at 7-10 days' interval as per soil moisture status during summer season. Watermelon cannot withstand water stagnation. Hence, over watering is to be avoided.

The pests like beetles and fruit fly are found in watermelon. The plant growth and yield is drastically affected due to infestation of these insects. The fruit flies enter the fruit at early stage and damage the fruit. These insects can be controlled by spraying of Malathion or profenophos @ 2 ml per litre of water at weekly intervals. Insects can also be controlled by spraying *neem* oil or neem based pesticide. Watermelon is sometimes affected by downy mildew disease. The plant growth is adversely affected due to infestation of this disease. The disease can be controlled by spraying of Mancozeb or Chlorothalonil 2 g/lit twice at 10 days' interval.

Fruits are ready for harvest when it produces dull sound upon tapping. The fruits are picked carefully from the plant and carried to shade. On an average, 6-10 kg of fruits can be harvested from each pit. However, the yield varies depending upon the variety, management practices and local climatic condition.

Ladies Finger

Ladies finger is an important vegetable crop of India. It is commonly known as okra or *bhindi*. It can be grown both in tropical and sub-tropical regions. It is a nutritious vegetable containing 1.9 % protein, 0.2 % fat, 6.4 % carbohydrate, 0.7 % minerals and 1.2 % fibres. During the growing period, it requires long warm season and high humid condition. The ideal temperature for growing okra is from 22 to 35°C. Temperature below 20°C hinders seed germination. It is highly sensitive to frost injury. It can be taken up in the pots, container or in small plots in urban areas. Pusa Makhmali, Punjab no.13, Parbhani Kranti,

Arka Anamika, Arka Abhaya, Azad Kranti, Punjab Padmini, Pusa Sawani, Varsha, Uphar, etc. are some of the commonly cultivated varieties of okra.

Ladies finger can be taken up in wide range of soils starting from sandy loam to clay loam. However, well drained sandy loam soil with high organic matter is suitable for growing okra. The optimum pH range is 6.0-6.8. It is normally grown during *kharif* and summer seasons. The *kharif* crop is sown during June-July and summer crop is sown during January-February.

The land is to be prepared properly to get a fine soil tilth. About 3-4 kg well decomposed farm yard manure or compost may be applied in each pit. In case of sowing in the kitchen garden about 25 kg well decomposed farm yard manure or compost is to be applied in an area of 10 m^2. Seed rate usually depends on the germination percentage, spacing and growing season. Seed treatment is done with *Tricoderma viridae* or *Pseudomonas fluorescens*. Alternatively, seeds can be treated with carbendazim or thiram to prevent spread of fungal diseases. Okra is sown either on flat beds or on ridges. During *kharif* season, seeds are sown at a spacing of 60 cm x 30 cm by accommodating about 55 plants per 10 m^2 area. During summer season, sowing is done at a spacing of 45 cm x 30 cm by accommodating about 74 plants per 10 m^2 area. The recommended fertilizer dose is 120 gram nitrogen, 60 gram phosphorus and 60 gram potash for a plot with 10 m^2 area. However, fertilizer should be applied based on soil test results. All the phosphorous and potash along with half of nitrogen is to be applied at the time of planting. The balance quantity of nitrogen is to be top dressed at 20-25 days after sowing. Application of neem or karanja oilcake at the time of sowing helps in improving the plant growth and obtaining higher yield. Application of *Azospirillum* and Phosphobacteria by mixing with FYM before sowing helps to get more yield. Foliar spray of boron and zinc helps in better plant growth and yield. The plots are to be kept weed free for a successful crop growth. Weeding is done manually by hoeing at 20-25 days after sowing followed by earthing up. During summer season, the crop requires adequate soil moisture for better plant growth. Irrigation is given at 7-10 days' interval as per soil moisture status. Okra cannot withstand water stagnation. Hence, care should be taken to drain out excess water from the field during rainy season.

Okra is affected by several diseases like yellow vein mosaic, powdery mildew, etc. In case of yellow vein mosaic, intervenal area becomes completely yellow or white. This disease is caused by a virus, being transmitted by the insect vector white fly. Due to this infection, flowering and fruit settings are restricted. Growing of resistant varieties such as Parbhani Kranti, Arka Abhay, Arka Anamika, etc. help to prevent this disease. Control of insect vector by spraying of profenophos or fenvalarate can control the disease infestation. The infestation of powdery mildew results in greyish powdery growth on the upper surface of the leaf causing reduction in fruit yield. It can be controlled by application of mancozeb or copper oxychloride.

Fruit borer is a major insect of okra. It feeds on the flowers and causes circular boreholes on fruits. Spraying of *Bacillus thuringiensis* @ 2 g/ litre water can control this insect. Aphid is another harmful insect sucking cell sap from the plant. Due to pest infestation, there is curling and crinkling of leaves resulting in stunted growth of the plant. It can be controlled by spraying of Imidachloprid or Dimethoate.

Okra crop starts harvesting at 55-65 days after planting when the pods are at tender stage. Delay in harvesting may deteriorate the quality. Picking of fruits is done after every 2-3 days. About 8-10 kg yield is obtained from a plot of 10 m^2 area. However, the yield varies depending upon the variety, management practices and local climatic condition.

Pea

Pea is primarily a cool season crop and cultivated during winter season in eastern and northern plains of India. However, in cooler areas located at higher altitude, it can be taken up round the year. The green seeds are used as vegetables and the dry seeds are used as pulses. Peas are low in calories, but rich in dietary fibre and protein. Pea also contains substantial quantity of vitamin A, vitamin B, vitamin C, vitamin K, zinc, potassium, magnesium, iron, etc. Peas help to strengthen immune system and reducing inflammation. Green plant is a good source of livestock fodder. Pea helps in enriching the soil fertility. The optimum temperature for pea cultivation ranges from 15 to 30 degree Celsius. Harvesting pea at 15 to 20 degree Celsius improves the quality and taste of the produce. It can be grown in the home garden in small patches. A few plants can also be grown in the pot or container placed on the rooftop of the house. Some of the important pea varieties are Early Bedger, Asauji, Jawahar Matar, Mithi Fali, Arkel, Alaska, Meteor, Lucknow Boniya, Little Marvel, Hisar Harit, Jawahar Matar-3, Jawahar Matar-4, Pant Matar 2,Pant Sabzi Matar-3, Kanwari, Bonneville, T-19, NP-29, Rachna, Pant Uphar, Arka Ajit, Punjab-88, VL Matar-3, etc.

Pea can be cultivated in a variety of soil. However, well drained loamy soil with pH ranging from 6.0 to 7.5 is suitable for this crop. Pea cannot thrive in waterlogged areas. Liming is to be done for acidic soil. The seed requirement is 50 to 75 gram for 10 m^2 area. Seeds are to be treated with carbendazim or thiram to prevent seed borne disease. Alternatively, *Trichoderma viridae* or *Pseudomonas fluorescens* can be used for seed treatment. Seeds are to be inoculated with *Rhizobium* bacteria culture before sowing to enhance nitrogen fixation. Sowing can be done during the month of September-October. In hilly tracts, sowing can be done during June-July.

Land preparation can be done thoroughly to obtain a good soil tilth. Application of 20-30 kg FYM or compost is to be done in an area of 10 m^2 for improving soil fertility. Seeds are sown in lines at a spacing of 30-45 cm x 10 cm depending on growing season and variety. Being a leguminous crop, pea requires lesser dose of nitrogenous fertilizer. The recommended dose is 60 gram nitrogen, 80 gram phosphorus and 70 gram potash as basal fertilizer for 10 m^2 area. About 60 gram nitrogen can be top-dressed at 3-4 weeks after sowing in an area of 10 m^2. Weeding should be done at three weeks after sowing to make the field clean. Application of boron and zinc may enhance the yield of the crop. Irrigation is given immediately after sowing for better germination. Subsequently, irrigation is to be given at an interval of 7-10 days.

Pea leaf miner is an important insect damaging the crop. It can be controlled by application of Dimethoate 30EC @ 2ml/ litre of water. Attack of thrips and aphids result in poor plant growth and yield. These insects can be managed by application of imidachloprid @ 2.5 ml with 10 litre water. Sometimes, pod borer infestation damages the produce. It can be controlled by spraying of fenvalarate. Insects can be controlled by use of yellow sticky trap or light traps. Pheromone trap can also be used to control insects. The crop is affected by powdery mildew disease. It can be controlled by application of wettable sulphur. Foliar spray of neem oil or neem based pesticide can also control the attack of insect and disease.

Harvesting is done just before maturity of the pods, when the colour of pod is green. Depending on the variety, harvesting starts at 45 to 60 days after sowing. About 6 to 7 kg pods can be obtained from an area of 10 m^2.

Cowpea

Cowpea is one of the most important leguminous vegetables grown across the country. It is cultivated for its long, green or purple colour pods used as vegetable. It is a warm season crop and grown mainly in tropical and subtropical areas. Warm humid climate is best suited for cultivation of vegetable cowpea. Cowpea can tolerate water stress condition to some extent, but susceptible to waterlogging. This crop grows well within a temperature range of 21-35 degree Celsius. Suitable temperature range for seed germination is 12-15^0 C. Cowpea cannot withstand frost. In urban areas, it can be taken up in small plots adjacent to the house. It can also be taken in pots and containers placed on the rooftop. It requires adequate sunlight for growth and development. The promising cowpea varieties are Arka Garima, Cowpea 263, Narendra Lobia 1, Pusa Barsati, Pusa Dofasali, Pusa Komal, Pusa Phalguni, Pusa Rituraj, Philippines Early, Yard Long Bean, etc.

Cowpea can be grown in different types of soil. However, well drained loamy soil or slightly heavy soil is suitable for cultivation. The crop cannot grow in saline or alkaline condition. This can be grown both during kharif and summer seasons. The kharif crop is sown during June–July and the summer crop is sown during the month of January- February. The seed requirement is 20-25 gram for 10 m^2 area. Seeds are to be treated with Carbendazim or thiram before sowing. Alternatively, *Trichoderma viride* or *Pseudomonas fluorescens* can be used for seed treatment. Seeds are to be inoculated with *Rhizobium* bacteria culture before sowing to enhance nitrogen fixation.

Land should be prepared thoroughly to obtain a good soil tilth. Farm yard manure or compost may be applied @ 15-20 kg in 10 m^2 area. Seeds are sown in lines at a spacing of 30-45 cm x 10-15 cm depending on growing season and variety. Being a leguminous crop cowpea requires lesser dose of nitrogenous fertilizer. The recommended dose is 20-25 gram of nitrogen and 50 gram each of phosphorous and potassium for a plot of 10 m^2 area. All the fertilizers are to be applied as basal dose. Irrigation is given immediately after sowing. Subsequently, irrigation is to be given at an interval of 7-10 days during summer season. Proper drainage is to be provided during kharif season to avoid water stagnation. The crop is kept free of weeds by hoeing and weeding three weeks after sowing. Pinching of terminal buds before flowering makes the plant bushy with more number of fruit bearing branches.

Aphid is the most important insect attacking cowpea. It is found on the tender branches and suck the cell sap. The plant growth is adversely affected due to attack of this insect. It can be controlled by spraying Imidachlorid or Dimethoate. Pod borer is another harmful insect damaging the pods. It can be controlled by foliar spraying of malathion or fenvalarate @ 2ml/lit. of water.

Powdery mildew disease causes considerable loss to the yield of cowpea. This can be controlled by spraying of mancozeb @ 3 g/litre of water. Sometimes, the plant is affected by yellow mosaic virus. It is a viral disease transmitted by white fly. There is appearance of yellow colour on the leaves and plant growth is affected. Control of whitefly by spraying of insecticide can prevent spread of this disease. Insect and disease can also be controlled by application of neem oil or neem based pesticides.

Harvesting starts from 45-50 days in the early cultivars, which may continue up to 100 days. Picking of pods for vegetable purpose is normally done at 15 days after pod setting. The pods are harvested at tender stage. Delay in harvesting may adversely affect the quality. On an average, the green pod yield is about 5-8 kg from an area of 10 m^2.

French Bean

French bean is a popular winter vegetable grown in India. The tender beans are very tasty and nutritious. It contains adequate quantity of carbohydrate, potassium and dietary fibre. It is a rich source of sodium, calcium, iron, vitamin A and vitamin C. It is a short duration crop mainly grown during rabi season. But, under hilly tracts, it can be grown during kharif and summer seasons. There should be adequate irrigation facility for growing this crop during rabi and summer seasons. This crop can be grown in small plots in the home garden or on the rooftop on a specially prepared bed with shallow soil. The promising varieties are Arka Komal, Premier, Arka Bold, Arka Sampoorna, Arka Suvidha, Arka Anoop, Arka Samrudhi, Arka Suman, Contender, Pant Anupama, Arka Karthik, etc. French bean is grown in cool climate with temperature range of 20^0 C to 25^0 C. The favourable soil temperature ranges from 18^0 C to 20^0 C for better root growth. It cannot tolerate frost condition.

French bean can be cultivated in wide range of soil. The most suitable soil is well drained loamy soil with pH range of 5.5-6.8. In warmer plain areas, the crop is taken during rabi season for which sowing is done during October-November. In areas at higher altitude, it can be grown during *kharif* and summer seasons. The sowing time for *kharif* crop is June-July and that of summer crop is February-March. The seed requirement is 60-80 gram for 10 m^2 area depending on variety, growing season and location. However, less seed is required for creeper type as compared to bushy type. Seeds are to be treated with Carbendazim or thiram to prevent infestation of diseases. Alternatively, *Trichoderma viride* or *Pseudomonas fluorescens* can be used for seed treatment. Seeds are to be inoculated with *Rhizobium* culture before sowing.

The land is to be prepared thoroughly for obtaining fine soil tilth. Stubbles of previous crops and perennial weeds are to be incorporated in the soil to improve the soil fertility. The soil should be enriched with addition of farm yard manure or compost @ 20 kg in 10 m^2 area at the time of land preparation. Seeds are to be sown in lines at a spacing of 45-60 cm x 20-30 cm. During *kharif* season, seeds can be sown on the ridges to protect the crop from waterlogging situation.

Chemical fertilizer plays an important role in obtaining good yield from french bean. Application of 40 gram nitrogen, 60 gram phosphorous and 50 gram potassium may be done in 10 m^2 area. All of phosphorous and potassium along with half of nitrogen are applied as basal dose. Rest of nitrogen is to be applied three weeks after planting. Foliar spraying of boron, zinc and other micronutrients enhances quality and quantity of the produce.

The critical periods of water requirement are germination, flowering and pod formation. In general, irrigation is given immediately after sowing. Thereafter, irrigation is to be given at an interval of 7-10 days depending on soil moisture availability. The crop is kept free of weeds by hoeing and weeding three weeks after sowing. Weeding should be followed by earthing up.

French bean is affected by various insects like aphids, thrips and pod borer. These insects suck the sap from the plant for which the plant growth is stunted. Spraying of imidachloprid or methyl demeton 25 EC or Dimethoate 30 EC can control the insects. Pod borer is another harmful insect damaging the pods. It can be controlled by foliar spray of malathion or fenvalarate @ 2 ml/litre of water. Insects can be controlled by use of yellow sticky trap, pheromone trap and light traps.

Among diseases, powdery mildew causes severe loss to the yield of french bean. This can be controlled by spraying of mancozeb or copper oxychloride @ 3 g/litre of water. Sometimes, the plant is affected by yellow mosaic virus. It is a viral disease transmitted by white fly. There is appearance of yellow colour on the leaves and plant growth is affected. Control of whitefly by spraying of insecticide can control this disease. Insect and pests can also be controlled by spraying of neem oil or neem based pesticides.

Depending on the variety and season of cultivation, the crop is ready for harvest about 40 to 50 days after sowing. The pods are picked when those are easily detached from the plant. Harvesting may be done during morning or evening hours to get good quality produce. The harvested produce is to be taken to shade immediately after harvesting. French bean can produce about 8-10 kg pods from an area of 10 m^2.

Cluster Bean

Cluster bean is otherwise known as guar. It is a drought tolerant warm season crop grown for tender fruits, which are used as vegetable. The pods contain about 10.8 % carbohydrate, 3.2 % protein and 1.4 % minerals. It is also a rich source of vitamin-A and vitamin-C. The plant is used as a nutritious fodder for livestock. Cluster bean, being a tropical plant, requires warm growing season.

The temperature ranging from 32 to 38^{0}C is suitable for vegetative growth. This crop grows well during summer and kharif season. It cannot tolerate shade and prefers long day condition for vegetative growth. Short day condition is suitable for flowering of the plant. This crop can be grown in the backyard garden in small patches or on the rooftop in specially prepared bed. Several varieties of cluster bean are grown in India. Some of those are Pusa Sadabahar, Pusa Navbahar, Pusa Mausumi, etc.

Cluster bean can be grown in all types of soils, but medium texture sandy loam soils are suitable for this crop. Cluster bean performs better in well-drained soil having pH value of 7.0 to 8.5. Waterlogging condition adversely affects the crop growth. It is grown during *kharif* and summer seasons. The *kharif* crop is sown during June–July. The summer crop is sown during January-February under irrigated condition. The seed requirement is 18-20 gram for 10 m^2 area. Seeds are to be treated with carbendazim or thiram to reduce infestation of fungal diseases. Alternatively, *Trichoderma viridae* or *Pseudomonas fluorescens* can be used for seed treatment. Seeds are to be inoculated with *Rhizobium* bacteria culture before sowing to increase nitrogen fixation.

Land is prepared thoroughly to obtain a good soil tilth. Application of farm yard manure or compost may be done @ 15-20 kg for 10 m^2 area at the time of land preparation. Seeds are sown at a spacing of 45 cm x 15 cm. Being a leguminous crop, it requires lesser dose of nitrogenous fertilizer. Basal application of N: P_2O_5:K_2O is to be done @ 25:50:25 gram for 10 m^2 area. Application of *Azospirillum* and Phosphobacteria enhances the yield. Intercultural operations may be done to keep the field free of weeds during initial stages. Irrigation is given immediately after sowing. The critical stages of water requirement are flowering and fruiting, where availability of soil moisture should be ensured. For obtaining higher yield, irrigation may be given at 7-10 days interval. Stacking with bamboo or wooden sticks prevents lodging in tall varieties and enhances production.

Aphid is the most important insect attacking cluster bean. It sucks the cell sap from tender branches and the plants become weak. It can be controlled by spraying of imidachloprid, dimethoate or methyl demeton. The crop is affected by powdery mildew and leaf spot diseases. These diseases can be controlled by spraying of carbendazim or mancozeb.

The crop is harvested at green pod stage. Delay in harvesting may adversely affect the quality. On an average, cluster bean yields about 4-6 kg pod from 10 m^2 area.

Radish

Radish is a root vegetable widely grown in tropical as well as cooler regions of the country. It is a crop of short duration and can be easily accommodated as intercrop with other vegetables. The edible roots are white or red in colour. Radish is primarily a cool season crop. It attains proper flavour and texture when grown under temperature regime of 15 to 20 °C. Long day and high temperature lead to poor root formation. Radish is more pungent at higher temperature and pungency decreases with decrease in temperature. It can be grown in small plots in the backyard garden or in pots and containers on the rooftop. There is large number of varieties cultivated in different regions of India. Some of the commonly grown radish varieties are Pusa Desi, Pusa Himani, Pusa Chetki, Pusa Reshmi, Arka Nishant, Japanese White, Punjab Safed, Kalyani White, Chinese Pink, etc.

Radish can be grown on nearly all types of soils, but the best yield is obtained in light loamy soil with high organic matter content. The ideal pH range is 5.5-6.8. Heavy soils are not suitable for this root crop. Radish requires thorough land preparation without clods for proper root development. The soil should not contain any undecomposed organic matter. Soil should be properly prepared to obtain fine soil tilth. Application of well decomposed farmyard manure may be done @ 20 kg in 10 m^2 area. Radish can be sown during September to November in plains and during June-July in the hills. In the regions with mild summer, it can be grown throughout the year.

About 10 gram seed is required for sowing in 10 m^2 area. Seed treatment may be done with carbendazim or thiram to protect the crop from seed borne diseases. Radish is usually sown in lines with a spacing of 30 cm X 15 cm. During rainy season, planting can be done on raised bed to avoid water stagnation. Fertilizer may be applied @ 50 gram nitrogen, 50 gram phosphorus and 50 gram potash in an area of 10 m^2. The full dose of potassium and phosphorous fertilizer along with half of nitrogen can be applied at the time of sowing. The remaining half of nitrogen fertilizer should be applied as top-dressing after three weeks of sowing. Spraying of boron and zinc helps in better plant growth and yield. Enough soil moisture should be available for uniform germination of seed and growth of the plant. The first irrigation is given immediately after sowing. Depending upon the season and the soil moisture availability, radish may be irrigated once a week. The root development is hampered in case of excess or deficit of moisture in the soil.

Regular weeding and hoeing are necessary to check the growth of weeds and keep soil in friable condition for better root growth. One hoeing followed by earthing up during the early stages of growth is necessary for proper development of roots. Radish root has a tendency to bulge out of soil as it grows in size. Hence, earthing up helps in covering the root to enhance the quality of produce.

Aphid is the most serious pest of radish. Humid weather favours the spread of aphid infestation. Leaves of the affected plants curl up, become yellowish and finally die. This can be controlled by spraying of imidachloprid @ 0.25 ml or dimethoate @ 2 ml /litre of water. Mustard saw fly is a common pest of radish. It appears when the crop is in flowering and vegetative stage. The insect makes holes in the leaves. This insect can be controlled by spraying quinalphos or profenophos @ 2 ml /litre of water.

Alternaria blight is a major disease affecting radish. The symptoms are yellowing of leaves and appearance of spots or lesions on the leaves. Infection spreads rapidly during kharif season. White rust is another disease of radish caused by fungal attack. There is appearance of white powdery substance in patches on the leaves. These diseases can be controlled by spraying fungicides like propiconazole or copper oxychloride.

Depending upon the cultivars, harvesting starts from 40-50 days after sowing. In case of backyard garden, the root yield of 20-30 kg can be obtained from 10 m^2 area.

Carrot

Carrot is an important root crop cultivated for its fleshy edible roots. It is used as raw material in food, pharmaceutical and cosmetic industry. Carrots are mostly consumed fresh, but many delicacies are also cooked from it. Most of the carrots are orange in colour. However, red, yellow or purple carrots are also found in different parts of the country. Carrot is a rich source of beta carotene, vitamin C, vitamin K, potassium, various minerals and dietary fibre. It helps in strengthening immune system and promotes digestive health. Carrot is a cool-season crop and grows well in the temperature range of 15° to 20°C. Carrot plant needs 6-8 hours of direct sunlight every day to form proper roots. This can be grown in small plots in the home garden. In urban areas, carrot can also be taken in the container or pot. A pot mixture can be prepared by mixing light soil, peat moss, coco-peat and compost for growing carrot. Some of the commonly cultivated carrot varieties are Early Nantes, Punjab Black Beauty, Punjab Carrot Red, Pusa Kesar, Pusa Meghali, Chantenay, India Gold, etc.

Carrots can be grown well in a wide variety of soils. However, it performs better in well drained loamy soil. The soil having pH ranging from 6.0 to 7.0 is suitable for carrot cultivation. Carrot is normally grown during winter season, for which sowing is done in the months of October-November. In hilly areas, this crop can also be grown during *kharif* season. Carrot can be grown throughout the year at higher elevation provided there is assured irrigation. The seed rate is about 5 gram for 10 m^2 area. Seeds are to be treated with carbendazim or thiram before planting. Alternatively, seeds can be treated with *Trichoderma viridae* to prevent seed borne diseases.

The field should have loose, friable, deep and well-drained soil for better germination and plant growth. A fine soil tilth helps in better root development and higher yield. Application of 20-25 kg farm yard manure or compost can be done in 10 m^2 area at the time of final ploughing. Carrot is sown in lines with a spacing of 30 cm X 15 cm. During *kharif* season, planting can be done on ridges to avoid water stagnation.

Fertilizer applications should be based on soil analysis. However, 60 gram nitrogen, 50 gram phosphorus and 110 gram potassium can be applied in 10 m^2 area. All the potassium and phosphorous along with half of nitrogen may be applied as basal fertilizer at the time of sowing. The remaining half of nitrogen should be top-dressed after three weeks of germination. Application of boron and zinc helps in better plant growth and yield. Adequate soil moisture should be available to help uniform seed germination and growth of the plant. Irrigation is to be given immediately after sowing. Depending upon the season and soil moisture availability, irrigation can be done at 7-10 days interval. Root development is adversely affected due to moisture stress. Excess moisture in soil results in small and light coloured carrots. Weeding should be done at early crop growth stage to check the growth of weeds. Earthing up is necessary for proper development of roots. Earthing up helps in covering the roots to get good quality carrot.

Carrot is affected by various diseases and insects. Sometimes, brown colour patches are seen on the leaves due to infestation of leaf spot disease. This disease can be controlled by spraying propiconazole or mancozeb. Alternaria blight is another disease of carrot. The symptoms are yellowing of leaves with appearance of raised spots or lesions on the leaf surface. It can be controlled by spraying fungicides like propiconazole or copper oxychloride.

Among insects, aphid cause serious problem. Cloudy weather and high humidity favour aphid infestation. Aphids suck cell sap from the leaves and tender parts of the plant. Leaves and shoots of the affected plants become yellowish. This can be controlled by spraying of imidachloprid @ 0.25 ml or dimethoate @ 2 ml /litre of water.

Carrots are harvested when the roots are fully developed. Harvesting should be done before full maturity. Delay in harvesting may result in hardiness of root, which become unfit for consumption. Carrot is harvested by uprooting manually or by digging with ordinary plough. The field needs to be provided with light irrigation a day before harvesting to facilitate uprooting. Depending on variety, growing season and management practices; about 25-30 kg of carrot can be obtained from 10 m^2 area.

Beet Root

Beetroot is a vegetable with blood red fleshy roots. It plays a vital role in maintaining human health by reducing blood pressure and improving digestion. It is very easy to cultivate beet root. Beet root can be grown all around the year under cool climatic conditions. However, over exposure to cold temperature can lead to slow growth of the plant. Beetroot loses its colour and quality under hot weather situation. The important varieties of beet root are Crimson Globe, Early Wonder, Detroit Dark Red, Ooty 1, Red Ball, etc. In urban ecosystem, beet root can be taken in small plots or in the pots. For the purpose, a pot mixture containing light soil, peat moss, coco-peat and compost can be used.

Beet root can grow well in various types of soils. However, well drained sandy loam or loamy soil with adequate quantity of organic matter is good for this crop. The suitable soil pH is 6.0-7.5. Beet root is normally grown in winter season for which sowing is done during October-November. The seed requirement is about 5 gram for 10 m^2 area. Seeds are to be treated with carbendazim or thiram before planting. Alternatively, seeds can be treated with *Trichoderma viridae* to prevent seed borne diseases. The soil should be properly prepared to get the desired yield. The field should have loose, friable and well-drained soil for better germination and plant growth. Application of 20-25 kg farm yard manure may be done in 10 m^2 area at the time of land preparation. Beet root is sown in lines with a spacing of 30-40 cm x 15-25 cm.

Fertilizer recommendation depends on results of soil analysis. However, application of 120 gram nitrogen, 160 gram phosphorus and 100 gram potassium is required for 10 m^2 area. All the potassium and phosphorous along with half of nitrogen is applied as basal fertilizer at the time of sowing. The remaining half of nitrogen should be top-dressed about 3-4 weeks after sowing. Application of boron helps in better plant growth and yield. Adequate soil moisture should be available to obtain uniform germination and better plant growth. Irrigation is to be given immediately after sowing. Depending upon the season and soil moisture availability, irrigation can be done once a week. Root development is adversely affected due to moisture deficit or excess moisture in soil. Weeding should be done at early growth stage to arrest the growth of weeds. Earthing up is to be done to facilitate irrigation.

Beet root is affected by fungal diseases like leaf spot disease. This disease can be controlled by spraying of propiconazole or carbendazim. Among insects, aphid causes serious problem for the plant. Humid weather condition favours aphid infestation. Aphids suck cell sap from the leaves and tender parts of the plant resulting in stunted plant growth. This can be controlled by spraying of imidachloprid @ 0.25 ml or dimethoate @ 2 ml/litre of water. Spraying of neem oil or neem based pesticide can be used for plant protection.

Beet root is harvested when the foliage start wilting. Delay in harvesting makes the root woody and unfit for consumption. Beet root is harvested by uprooting manually. There should be adequate moisture in the field to facilitate harvesting. In home gardens, about 15-20 kg of beet root can be obtained from 10 m^2 area.

Cabbage

Cabbage is a cool season crop consumed as leafy vegetable. It is one of the most popular and widely grown vegetables in the country. The green or purple coloured head is a rich source of vitamin A, vitamin C and minerals like phosphorus, potassium, calcium, sodium & iron. Cabbage can be eaten as salad or cooked form. In India, cabbage is mainly grown in winter season in the plains and round the year in hilly tracts. Cabbage requires cool and moist climate for better head formation. The suitable temperature range for seed germination is 13-16^{0}C. The growth is affected at temperature above 25^0 C. The typical sweet smell of cabbage gets deteriorated under dry and warm weather condition. It can be grown in small plots in the backyard or in the pots/containers placed on the rooftop or terrace. The predominant varieties are Green Express, Pusa Ageti, Pride of India, Green Challenger, Green Hero, Rare Ball, Golden Acre, Pusa Drumhead, Pusa Mukta, etc.

Cabbage performs well in loamy to sandy loam soil with high quantity of organic matter. However, it can be grown in clay loam or silty soils. Early crops prefer light soil while late season crops thrive better on heavier soils due to higher water holding capacity. A pH range of 5.5 to 6.5 is suitable for obtaining higher production. The seeds are to be treated with thiram or carbendazim to get rid of fungal diseases. Although this is a rabi season crop, it can be grown during kharif season in hilly tracts. The winter season crop is sown during October-November and *kharif* crop is sown during June-July.

Cabbage seedlings are transplanted for propagation. Seedlings are prepared in raised nursery beds. The beds should be properly prepared by adding well rotten FYM or compost @ 4 kg per m^2 area. The seeds are sown in lines at a spacing of 8-10 cm. The bed is covered with mixture of fine soil and powdered

FYM after sowing of seeds. Thereafter, light irrigation is provided with rose can. The nursery bed should be kept weed free.

Proper field preparation is done to make the soil friable. FYM or compost is incorporated in the soil @ 15 to 20 kg per 10 m^2 during land preparation. The seedlings are transplanted at 3-4 weeks stage having 4-6 leaves. The seedlings are transplanted at a spacing of 45-60 cm x 45 cm depending on the varieties. Transplanting should be done preferably in the evening hours. Irrigation should be provided just after transplanting for better survival.

Cabbage requires adequate nutrition for better plant growth and yield. Application of 120 gram nitrogen, 60 gram phosphorus and 60 gram potash is to be done in an area of 10 m^2. All the phosphorus and potash along with half of nitrogen is applied as basal fertilizer at the time of transplanting. Balance quantity of nitrogen is top-dressed in two equal splits at three and five weeks after transplanting. Application of boron plays a vital role in enhancing the yield and quality of cabbage. Weeding is necessary to keep the field clean. It should be followed by earthing up to facilitate irrigation and proper drainage. In urban areas, cabbage can be grown in small plots or even in the pots. It is required to provide adequate quantity of organic manure in the forms of FYM, vermi-compost or oilcake to get good yield from the pots. Application of higher quantity of chemical fertilizer in the pots may be harmful for the plants grown in the pots.

Cabbage suffers from various insects and diseases. Cutworms cause damage by cutting the foliage and young seedlings. It can be managed by picking and destruction of the larvae at the early stage of the crop. Application of profenophos @ 2 ml/litre water can control the insect. Diamond backmoth is another harmful insect causing damage in cabbage plant. Spraying of Cartap hydrochloride @ 1 g/litre or *Bacillus thuringiensis* @ 2 g/litre water can control this insect. Installation of pheromone trap or yellow sticky trap and spraying of neem oil can also control the insect attack.

Damping off is a serious disease found in the nursery. Due to attack of this disease, the affected seedlings fall off at the collar region. This can be controlled by spraying of propiconazole or copper oxychloride. There is appearance of chlorotic patches on the leaf due to infestation of black rot. Symptoms may appear from any side or centre of the leaves. This can be controlled by two sprayings of neem oil or neem based pesticide.

Harvesting is done when the heads are well developed and firm. The heads are cut with a knife and taken to shade. From home gardens, the yield of 25-30 kg can be obtained from an area of 10 m^2.

Cauliflower

Cauliflower is an important winter vegetable grown across the country. It is a rich source of carbohydrate, protein, vitamin-B, vitamin C, fat, sodium, potassium, etc. Cauliflower is a preferred vegetable due to its attractive appearance, good taste and nutritive value. Cauliflower is grown for its edible flowering head and consumed as a vegetable in curries, soups and pickles. It requires cool moist climate. The temperature ranging from 17 to 23 °C is suitable for this crop. The early varieties may tolerate higher temperature. This can be grown in plains during winter months, but can be cultivated round the year in higher altitudes. In urban areas, it can be grown in small plots in the backyard area or in the pot/container placed on the rooftop or terrace. Cauliflower varieties are of three categories depending on season of growing. The early season varieties are Early Kunwari, Early Synthetic, Pusa Katki, Pant Gobhi-2, Pant Gobhi-3, etc. Some of the main season varieties are Pusa Synthetic, Pant Shubhra, Punjab Giant-26, Punjab Giant-35, etc. Pusa Snowball-1, Pusa Snowball-2, Sonwball-16, Dania Kalimpong, etc. are suitable for late season crop.

Cauliflower can be cultivated in various types of soil starting from clay loam to loamy soil. However, deep loamy soil with high moistureholding capacity is suitable for the crop. Water stress condition adversely affects curd formation. For rainy season crop, well drained soils are preferred as the crop cannot tolerate water stagnation. Cauliflower is sensitive to high soil acidity and moderately sensitive to salinity.

Cauliflower can be sown from September to October and are ready for harvest during December-January. However, time of sowing may vary depending on variety and local climatic condition. The seeds should be treated with thiram or carbendazim to get rid of fungal infection.

Seedlings are prepared in raised nursery beds. FYM or compost is added in the nursery bed @ 4 kg per m^2 area. The seeds are sown in line at a spacing of 8-10 cm. After sowing of seeds, the bed is covered with the mixture of fine soil and farm yard manure. Thereafter, light irrigation is provided to ensure germination. During monsoon, nursery beds should be prepared under polytunnel to protect from heavy rain. Weeding and intercultural operation should be done as and when required. Nursery beds should always be irrigated according to need.

Field preparation is to be done properly to get fine soil tilth. FYM or compost is incorporated in the soil @ 15 to 20 kg per 10 m^2 area during land preparation. The seedlings are ready for transplanting at 3-4 weeks after sowing. The seedlings are transplanted at a spacing of 45-60 cm x 45 cm depending on the varieties and season. Transplanting should be done in the evening hours for better establishment. Irrigation should be provided just after transplanting.

Cauliflower requires 120 gram nitrogen, 60 gram phosphorus and 60 gram potash per 10 m^2 area for better growth and yield. All the phosphorus and potash along with half of nitrogen are applied as basal fertilizer at the time of transplanting. Rest quantity of nitrogen is top-dressed in two equal splits at three and five weeks after transplanting. Application of boron plays a vital role in enhancing the yield and quality of cauliflower. Weeding is necessary to keep the field clean. It should be followed by earthing up to facilitate irrigation and drainage. In urban areas, cauliflower can be grown in small plots or even in the pots. Adequate quantities of organic manure in the forms of FYM, vermi-compost or oilcake are to be provided to obtain better plant growth and yield from the potted plants. Application of higher quantity of chemical fertilizer in the pots may be detrimental for the plants grown in the pots. Exposure of the curds to bright sunlight may result in loss of flavours and colour. The curds are to be covered by tying the tips of the outer leaves with twine or rubber band to retain desired colour and flavour of the curd.

Cauliflower suffers from various insects and diseases. Cutworm causes damage by cutting the foliage and young seedlings. It can be managed by picking and destruction of the larvae at the early stage of the crop. Application of profenophos @ 2 ml/litre of water can control the insect. Diamond back moth is another harmful insect causing damage to cauliflower plant. Spraying of Cartap hydrochloride @ 1 g/litre of water or *Bacillus thuringiensis* @ 2 g/litre of water can control this insect. Installation of pheromone trap & yellow sticky trap and foliar spray of neem oil or neem based pesticides can control the insect attack.

Damping off is a serious disease found in the nursery bed. Due to disease infestation, the affected seedlings droop down and fall off at the collar region. This can be controlled by spraying of propiconazole or copper oxychloride. There is appearance of chlorotic patches on the leaf due to infestation of black rot. Symptoms may appear from any side or centre of the leaves. This can be controlled by spraying of propiconazole or mancozeb.

Cauliflower suffers from various physiological disorders due to deficiency of several micronutrients or growing under disadvantage situation. Browning or brown rot is caused due to boron deficiency. In this case, water soaked areas appear on the curd and later changes into rusty brown. Basal application or spraying of boron can address the problem. Whip tail is caused due to deficiency of molybdenum. It is more pronounced in acidic soil. The leaf blades do not develop properly due to whip tail. It can be managed by spraying of sodium molybdate at 30 days after planting. Nitrogen deficiency and delayed planting may cause buttoning. There is development of small curds or buttons due to this problem. In this case, the leaves remain small and do not cover the developing curds. Application of recommended quantity of nitrogen and avoiding transplanting of over-aged seedlings can solve this problem.

Harvesting is done when the heads are well developed and firm. Delay in harvesting results in discolouration and deformation of the curd. Therefore, it is essential to harvest cauliflower at right maturity stage. The curds are harvested by cutting with a knife. In case of home garden, about 25-30 kg yield can be obtained from an area of 10 m^2.

Broccoli

Broccoli is an exotic vegetable gaining popularity in India for its taste and health benefits. It is consumed as salad and curry. Broccoli is a rich source of vitamins and minerals. It also contains antioxidants. Broccoli crop requires cold weather and can be cultivated during winter season. The ideal temperature for raising a successful broccoli crop is 15-23° C. However, it can be cultivated round the year under control condition in green house. In urban areas, broccoli can be grown in small plots or pots. It is required to provide adequate quantity of organic manure in the forms of FYM, vermi-compost or oilcake to get good yield from the potted plants. Application of higher quantity of chemical fertilizer in the pots may be detrimental for the plants. Some of the popular broccoli varieties cultivated in India are Evergreen, Danube, Green Mountain, Pusa Broccoli, KTS-I, Solan Green, Lucky, Fiesta, Pushpa, Aishwarya, Palam Vichitra, etc.

Broccoli can be grown in a wide range of soil. However, sandy and silty loam soils are preferred to obtain good yield. The ideal soil pH is 5.5 to 6.5. Broccoli is suitably grown during winter season and the sowing time is October to November. However, time of sowing may vary depending on variety and local climatic condition. Broccoli is propagated by seedlings. Seedlings are grown in raised nursery beds The seeds should be treated with thiram or carbendazim to prevent infestation of fungal diseases. The nursery beds should be properly prepared by adding well rotten FYM or compost @ 4 kg/m^2. The seeds are sown in line at a spacing of 8-10 cm. After sowing, the bed is covered with mixture of fine soil and powdered FYM. Light irrigation is provided after sowing to get good germination of seed. Weeding should be done in nursery and irrigation is done as per need of the soil.

Proper preparation of main field should be done to get fine soil tilth. FYM or compost is to be incorporated in the soil @ 15 to 20 kg for 10 m^2 area during land preparation. The seedlings are ready for transplanting at 3-4 weeks stage. The seedlings are transplanted at a spacing of 45-60 cm x 45 cm depending on the varieties. Transplanting should preferably be done in the evening hours. Irrigation should be provided just after transplanting for better survival of seedlings.

Fertilizer should be applied based on soil test results. However, broccoli requires 120 gram nitrogen, 60 gram phosphorus and 60 gram potash in 10 m^2 area for better yield. All the phosphorus and potash along with half of nitrogen is applied as basal fertilizer at the time of transplanting. Balance quantity of nitrogen is top-dressed in two equal splits at three and five weeks after transplanting. Application of boron plays a vital role in enhancing the yield and quality. Weeding is necessary to keep the field clean. It should be followed by earthing up for better plant growth.

Broccoli is affected by various insect and diseases. Diamond back moth is a harmful insect causing damage in broccoli plant. Cutworms also cause damage by cutting the foliage and young seedlings. It can be managed by picking and destruction of the larvae at the early stage of the crop. Application of profenophos @ 2 ml/litre of water can control these insects. Installation of pheromone trap&yellow sticky trap and spraying of neem oil or neem based pesticide can control the insect attack.

Damping off is a serious disease found in the nursery bed. In case of disease infestation, the affected seedlings are damaged. High humidity condition, heavy rainfall, poor drainage condition and low temperature are favourable for this disease. This can be controlled by spraying of propiconazole @ 2 ml/litre of water. There is appearance of chlorotic patches on the leaf due to infestation of black rot. The plants wilt and die if this infection is observed at early stage of growth. This can be controlled by foliar spray of neem oil or neem based pesticide.

Harvesting is done when the heads are well developed. Delay in harvesting results in deformation of the curd, which diminishes market demand. Therefore, it is essential to harvest the cauliflower at the right maturity stage. From backyard cultivation, 20-25 kg yield can be obtained from an area of 10 m^2.

Colocasia

Colocasia, commonly known as *arbi* or taro is a starchy vegetable. All the plant parts of colocasia are edible. The leaves, petioles and corms are used for preparation of various dishes. It is a rich source of carbohydrate, minerals and vitamins. It is a crop of tropical and sub-tropical regions and requires warm humid climate for better growth and yield. The ideal temperature range for colocasia cultivation varies from 21 to 27^0 C. It can be planted in both *kharif* and summer seasons. The *kharif* crop is planted during June –July and summer season crop is planted during February-March. It can be grown in the backyard to fulfil the family requirement. Some of the promising varieties are Muktakeshi, Panchamukhi, Satamukhi, Sree Pallavi, Sree Kiran and Sree Rashmi.

Colocasia grows well in sandy loam or alluvial soil with abundant organic matter content and good moisture holding capacity. A soil pH of 5.5-6.5 is ideal for this crop. Cormels or side tubers weighing about 25-30 g are used as planting material. Being a tuber crop, colocasia requires fine soil tilth for better growth of the corms. The soil is thoroughly prepared to obtain desired soil tilth. Application of farm yard manure or compost @ 20-25 kg per 10 m^2 area is done during final land preparation. Ridges and furrows are laid out at a spacing of 60 cm. The cormels are planted at a spacing of 45 cm in the furrows. To ensure uniform sprouting, irrigation should be given just after planting and repeated one week after planting. Subsequently, water is applied at 12-15 days' interval depending on the soil moisture status. Irrigation is stopped one month prior to harvest to facilitate easy harvest and drying of the tubers. Taro requires about 80 gram nitrogen, 60 gram phosphorus and 80 gram potash per 10 m^2 area. Half of nitrogen and potassium along with full of phosphorus is applied as basal dose. Rest of the nitrogen and potassium are applied during first hoeing and earthing up. Some persons prefer to grow colocasia by using organic matters like well decomposed farm yard manure, compost, vermi-compost, oilcake, etc. Weeding, light hoeing and earthing up is required at 25-30 days and 60-75 days after planting for weed control and loosening the soil. New suckers arising from the mother plants have to be removed during second weeding. The leafy parts may be smothered about one month before harvest so as to enhance tuber development.

Colocasia is affected by leaf blight disease. Due to this disease, irregular brown coloured necrotic patches appear on leaves. This can be controlled by spraying fungicides like propiconazole or mancozeb.

Colocasia is generally harvested after drying of the leaves. Early harvesting may reduce the yield. In backyard cultivation, about 12-15 kg yield can be obtained from 10 m^2 area.

Dioscorea

Dioscorea or yam is a popular tuber crop of India. It requires warm and humid conditions with a mean temperature of 25 -35^0 C. It cannot withstand waterlogging situation. The crop is susceptible to frost. It is a good source of carbohydrate, protein, vitamins and minerals. It is used for preparation of fry, chips and flakes. It can be planted along the boundaries of home gardens in urban areas. The popular varieties include Sree Roopa, Sree Keerthi, Sree Shilpa, Siruvalli, Sree Latha, Sree Kala, Arka Upkar, Odisha Elite, Sree Priya, etc.

Dioscorea can be grown in various types of soil. However, sandy loam or red loamy soil with pH ranging from 6.0 to 6.5 is suitable for this crop. The soil should be rich in organic matter content and have good drainage facility. This is normally grown during kharif season. The best planting time is May-June. The tuber cuttings with dormant bud or eye are used for propagation of yam. The larger size tuber cuttings result in early emergence of plant. The cut tubers are dipped in cow dung slurry or fungicide before planting to prevent infestation of seed borne disease.

Yam requires fine soil tilth for better growth of the tubers. Proper soil preparation is necessary to obtain desired soil tilth, which help better growth of the roots. Yam is planted on pits or mounts prepared at a spacing of 1 m x 1m. About 10,000 plants can be accommodated in one hectare area. FYM or compost is applied @ 3-4 kg per pit before planting. In some places, planting is done at a spacing of 75 cm x 75 cm. Mulching is advisable to obtain early sprouting. The mulching material also adds organic matter to the soil. To ensure uniform sprouting, irrigation should be given just after planting. Subsequently, water is applied depending on the soil moisture requirement. During kharif season, adequate care is to be taken for providing proper drainage. Yam requires about 10 gram nitrogen, 6 gram phosphorus and 8 gram potash per pit. Half of nitrogen along with full dose of phosphorus and potassium is applied as basal fertilizer. Rest of the nitrogen is applied after 5-6 weeks after planting. Application of *Azospirillum* at one month after planting is beneficial for plant growth and yield. Weeding should be done at 5-6 weeks after planting for better plant growth. Hoeing and earthing up at this stage kill the weeds and make the soil loose, which facilitates better tuber growth. The vines are to be staked on bamboo poles of 3-4 m length for better plant growth. The yield from staked plants is normally higher than other plants.

Sometimes, yam scale insect causes significant loss to the crop. Yam scale is found on the tubers both under field and storage conditions. Dipping the planting material in Methyl demeton 0.25% prevents spread of this insect. Scale free seed tubers should be used for planting. The plant is also affected by powdery mildew. The symptoms of this disease are presence of black, grey, white or pink coloured dust particle on the plant. The affected foliages are to be removed. There is appearance of dark brown, pin head like spots on the leaves and stems due to attack of anthracnose disease. Spraying of fungicides like carbendazim, propiconazole, mancozeb, etc. can control these diseases.

Yam becomes ready for harvest at 9-10 months after planting. It is harvested when the leaves turn yellow and start drying. The tubers are carefully harvested by digging. Injury to the tuber may reduce the market price. If harvesting is not done, it becomes a perennial plant. The stem will die at the end of the growing season and will give rise to a new stem in the subsequent year. A good crop can yield about 2.5-3.0 kg tuber per plant.

Onion

Onion is one of the most important commercial crops of India. It is used both as condiment and vegetable. Onion is consumed as raw and used during preparation of soup, sauce, etc. It can also be processed into dry products such as flakes and powder.It is grown under a wide range of climatic conditions. However, it cannot withstand extreme hot or cold weather conditions. The ideal temperature range for growing onion is 15-25^0 C. Short days are very much favourable for formation of bulbs. It can be grown well up to an altitude of 1300 meter above mean sea level. It can be taken in small beds in the backyard for home consumption. Some of the important varieties are Pusa Red, Pusa Ratnar, Pusa White Round, Arka Pragati, Patna White, Bombay White, Nasik Red, Bellary Red, NP 53, Arka Niketan, Arka Kalyan, Arka Lalima, Arka Kirthiman, Arka Pitamber, Bhima Super, Arka Kalyan, Agri Found Light Red, Agri Found Dark Red, etc.

Onion can be taken up in well drained loamy soil to get good yield. The soil with high organic matter and potassium content is suitable for this crop. An ideal soil should have pH in between 5.8 to 6.5 with good aeration. Heavy soil should be avoided for growing onion as it hinders the bulb formation. Onion is planted by raising seedlings in the nursery bed. Rabi season crop is sown during October to November. August -September is the ideal time for sowing of kharif crop. In hilly regions, the seeds can be sown during March-April. The seed rate is 10-15 gram for 10 m^2 area. Seeds are treated with carbendazim or thiram to prevent spread of soil borne diseases. Seeds are sown in raised nursery bed of three meter length one meter width. The beds are raised to a height of 15-25 centimetre for better drainage. There should be 60-70 centimetre gap between beds for facilitating cultural operations. Seeds are sown in lines with a distance of 4-5 centimetre. The seeds are to be covered with straw for better germination. The top layer of nursery bed is to be prepared with well decomposed farm yard manure. Seedlings are ready for transplanting at 6-7 weeks after sowing. The field is to be properly prepared to get a fine soil tilth.

Application of FYM or compost is to be done to get good yield. Seedlings are transplanted at a spacing of 15 cm x 10 cm.

A fertilizer dose of 150 gram nitrogen, 80 gram phosphorus and 60 gram potash is required for 10 m^2 area. All the phosphorus & potash and half of nitrogen are to be applied as basal dose. Rest of the nitrogen is to be top-dressed at 20-25 days after transplanting. Application of *Azospirillum* and phospho-bacteria along with farm yard manure enhances soil fertility. Neem or karanja oilcake may be applied before transplanting to minimize effect of soil borne diseases. Application of sulphur enhances the yield and improves the quality of onion. Hoeing is done at three weeks after transplanting to keep the field free from weeds. In urban areas, onion can be grown in small plots to meet the household demand. It is required to provide adequate quantity of organic manure in the forms of FYM, vermi-compost or oilcake to obtain good yield. Irrigation should be provided at regular intervals for better plant growth. Water stress may adversely affect bulb formation. Irrigation is to be discontinued one week before harvesting.

Onion is affected by various insects and diseases. Attack of insects like thrips and onion fly affects the plant growth seriously. These insects can be controlled by spraying Methyl demeton or Dimethoate @ 2 ml/litre water. Diseases like leaf spot and powdery mildew cause severe loss to onion crop. These fungal diseases can be controlled by spraying of propiconazole or copper oxychloride.

Onion is harvested when 75 per cent of the leaves are dried. The crop gives an yield of 10 to 15 kg from an area of 10 m^2 area. Bulbs are to be properly cured before storage.

Garlic

Garlic is an important spice crop grown in India. Besides, it has several medicinal properties and useful against stomach disease & common cold. It is grown under a wide range of climatic conditions. However, it cannot withstand too hot or too cold weather. It prefers moderate temperature for growth. Short days are favourable for growth and development of this crop. It can be grown up to an elevation of 1300 meter above mean sea level. In urban areas, it can be grown in small plots in the backyard or any place available near the building. The varieties of garlic are either white or of different colours with variation in pungency. Some of the improved varieties are Bhima Omkar, Bhima Purple, Yamuna Safed, Yamuna Safed-2, Shweta, VL Garlic 1, Ooty 1, etc.

Garlic requires well-drained, sandy loam soil with good moisture holding capacity for best of its growth and yield. The ideal soil pH for garlic crop is 6.0 - 7.0. Garlic responds well to high soil organic matter content. Garlic is propagated by cloves. All the cloves can be used as planting material, except the central clove. Healthy cloves, free from disease, should be used for planting. The cloves are to be treated with mancozeb @ 3 gm/litre of water for 30 minutes to protect from fungal infestation.

Planting is done during October-November in plain areas and during June-July in hilly areas. Soil preparation is done to get good tilth. Application of 20-25 kg of well decomposed farm yard manure or compost per 10 m^2 area is to be done at the time of land preparation to improve soil fertility. Planting is done either by dibbling or furrow planting. For planting, cloves are dibbled at a spacing of 15 cm X 7.5 cm at a depth of 5 to 7.5 cm with the growing end upwards. In case of furrow planting, the furrows are prepared at a distance of 15 cm and the cloves are planted in the furrow at a spacing of 7.5 to 10 cm. Those are covered with loose soil followed by light irrigation. The fertilizer requirement is decided by inherent fertility status of the soil. However, it is required to apply 120 gram N, 60 gram P_2O_5 and 120 gram K_2O in 10 m^2 area. All the phosphorus and half of nitrogen & potash are applied at the time of planting as basal fertilizer. After three weeks of planting, 30 gram of nitrogen is top-dressed. At six weeks after planting, rest of nitrogen and potash are to be applied.

Timely weed control is necessary for better plant growth and obtaining higher yield. Hoeing should be done at three weeks after planting to control the weeds. Hoeing at a later stage may damage the stem and hamper plant growth. Water management plays an important role in obtaining good yield. Irrigation is to be provided at 10 to 15 days' interval depending upon the soil moisture availability. Moisture scarcity during active growth stage adversely affects development of the bulbs. Irrigation should be stopped one week before harvesting for easy harvest. Proper drainage may be provided during kharif season to avoid water stagnation.

Garlic is affected by various insects and diseases. Attack of insects like thrips and aphids affects the plant growth seriously. These insects can be controlled by spraying of imidachloprid, methyl demeton or Dimethoate. Diseases like leaf spot and powdery mildew affect garlic crop. These fungal diseases can be controlled by spraying of propiconazole or copper oxychloride.

Garlic is harvested after the leaves turn yellow and dries up. The plants are pulled out or dug with spade for harvesting. Those are tied into small bundles and dried in shade for 2-3 days for curing to enhance the keeping quality. About 5 to 6 kg garlic can be harvested from a small plot of 10 m^2 area.

Ginger

Ginger is the most important spice crop of India. It performs well in humid condition. Ginger is suitable for intercropping in fruit orchards. Ginger can be cultivated in areas receiving abundant rainfall. The rhizome is used as a spice. Ginger can be used fresh, dried, powdered or as an oil or juice. It is an ingredient in various recipes and normally mixed with processed foods. Ginger grows well in the temperature range of 25 to 35 ^{0}C. In urban areas, ginger can be grown in the backyards or under the tree. If grown in the backyard, about 20 kg ginger can be harvested from 10 m^2 area. This will be adequate to meet the family requirement for one year. There are several popular varieties grown in India. Some of those are IISR Varada, IISR Mahima, Karthika, Suprabha, Suruchi, etc.

Ginger grows well in well drained sandy loam or red loamy soil. It cannot survive in waterlogged conditions. A pH range of 5.5-6.5 is required for good growth of crop. Ginger crop should not be grown in the same land continuously for year after another. Usually, ginger is a kharif season crop. Hence, planting is to be done during first week of May to June. However, there should be irrigation facility for planting the crop during May. Rhizomes are used for cultivation of ginger. The seed rhizomes should be fresh and free of disease infestation. About 15-16 kg of rhizome is required for planting in 10 m^2 land area. The rhizomes are to be treated with mancozeb @3 gm/ltr of water for 30 minutes to protect the crop from fungal infestation. After treatment, the rhizomes are to be dried in shade for 3-4 hours.

Field preparation should be properly done to get fine soil tilth. About 25-30 kg of FYM or compost may be applied in 10 m^2 area to improve soil fertility. Ginger is planted on raised beds of 15 cm height, one meter width and convenient length. There should be gap of 50 cm between the beds to facilitate field operation. Ginger is planted at a spacing of 30 cm x 20-25 cm on the raised beds. Mulching can be done with green leaves after planting to conserve soil moisture and promote quick germination. Ginger is a heavy feeder requiring higher quantity of plant nutrients. Basal application of 100 gram phosphorus and 50 gram potash is required per 10 m^2 area. It is required to apply 60 gram nitrogen in an area of 10 m^2at six weeks after planting. There should be application of 60 gram nitrogen and 50 gram potash in the plot of 10 m^2at three months after planting.

Weeds create problem in plant growth and reduces yield. Timely weed control is necessary for better plant growth. Hoeing may be done at three weeks after planting to control the weeds. Earthing up is required to enhance root development and for facilitating good drainage. In many places, ginger is grown as a rainfed crop. However, irrigation may be given at the time of need for better plant growth. Proper drainage is to be done during monsoon months to avoid water stagnation.

Ginger suffers from various types of disease and insect attack. It is affected by the fungal disease rhizome rot. Due to this disease, the leaves become yellow and die. This can be controlled by soil drenching with Mancozeb @3 gm/litre or Metalaxyl@1.25 gm/litre at 30, 60 and 90 days after planting.

Among insects, shoot borer is harmful for this crop. The insect enters in the stem after which plants start wilting. Spraying of neem oil or neem based pesticide can control this insect. Alternatively, spraying of dimethoate or profenophos @2 ml/litre water can control this insect.

The crop is ready for harvest after eight months of planting. Ginger is harvested when the leaves become yellow and dry out completely. The rhizomes are harvested by digging from soil. After harvesting, those are to be cleaned and dried under shade for 2-3 days. On an average, 15 to 17 kilogram of ginger can be obtained from 10 m^2 land area.

Turmeric

Turmeric is a popular spice, which contains a chemical called curcumin. Turmeric has a long history of medicinal use. Turmeric is known for its medicinal value and used as an antiseptic in India since ancient times. It is used not only as a spice, but also as an important component in religious ceremonies. Turmeric has a bitter taste and is often used in curry for flavour or colour. It is widely used in dye, drug and cosmetic industries. Turmeric needs a warm and humid climate for growth. It grows well in hilly areas at an altitude of 1500m from mean sea level. The ideal temperature range between 20-35 ^{0}C is suitable for turmeric cultivation. It can be grown under rainfed and irrigated conditions. In urban ecosystem, it can be grown in backyard with minimum use of land. It can also be accommodated under the tall trees like coconut, sapota, guava, mango, etc. Some of the popular high yielding varieties of turmeric are Suguna, Krishna, Sudarshana, Sugandham, Roma, Ranga, IISR Prabha, IISR –Pragati, IISR-Kedaram, Suvarna and Lakadong.

Turmeric requires fertile and friable soil for better growth. It grows well in properly drained sandy loam soil with high organic matter content. It can also be grown in different types of soil such as sandy loam, light black, clay loam or red soil. Excess alkaline or acidic soil adversely affects growth of turmeric plant. Turmeric is usually grown during kharif season. Planting is to be done during first week of May to June. However, there should be irrigation facility for planting the crop during May. Rhizomes are used for cultivation of turmeric. The seed rhizomes should be free from disease pathogen. About 20-25 kilogram of rhizome is required for planting in 10 m^2 area. The rhizomes are to be treated with mancozeb @3 gm/litre of water for 30 minutes to protect the crop from fungal infestation. After treatment, the rhizomes are to be dried in shade for 3-4 hours.

Field preparation is done to get fine tilth of soil. Application of 20-25 kilogram of well decomposed farm yard manure or compost can be done in an area of 10 m^2 at the time of land preparation. Turmeric is planted in raised beds of one meter width and 15 cm height. There should be a gap of 50 cm between the beds to facilitate field operation. In urban areas, soil solarisation can be done before planting. In this process, the soil is ploughed, moistened and covered with a polythene sheet. The polythene sheet is removed after 15-20 days. It helps in keeping the soil free from weeds, insects and pathogens. Solarisation of beds will prevent spread of various soil borne diseases.

The rhizomes are planted at a spacing of 30 cm x 20-25 cm on the raised bed. Mulching can be done with green leaves or crop residues after planting to conserve soil moisture. Mulching also helps in better germination. Turmeric crop is a heavy feeder requiring higher quantity of plant nutrients. The fertilizer requirement is adjudged by inherent fertility status of the soil. However, it is recommended to apply 60 gram nitrogen, 30 gram phosphorus and 90 gram potash in 10 m^2 area. All the phosphorus and half of potash are applied at the time of planting as basal fertilizer. Nitrogen is applied in two equal splits at six weeks and twelve weeks after planting. Half of potash is applied at twelve weeks after planting.

Timely weed control is necessary for better plant growth and higher yield. Hoeing should be done at three weeks after planting to control the weeds. Earthing up is required for better drainage and root development. Water management plays an important role in obtaining good yield. Irrigation may be given at the time of need to maintain growth of the plant. Proper drainage may be provided during kharif season to avoid water stagnation.

Turmeric suffers from attack of various disease and insect. Rhizome rot is an important fungal disease affecting the crop. Due to disease infestation, the leaves become yellow and die. This can be controlled by soil drenching with copper oxychloride or mancozeb or metalaxyl. Among insects, shoot borer is harmful for this crop. The larvae of the insect burrow into the pseudo-stems and feed on the growing shoot. Spraying of neem oil or neem based pesticide can control this insect. Alternatively, spraying of dimethoate or profenophos @2 ml/litre water can be done to control this insect. Yellow stick trap or light trap can be installed to check the insect attack.

Depending on the variety and local climatic condition, turmeric is ready for harvest within 7-9 months of planting. The crop is ready for harvest when the leaves and stem start turning brown and dry up progressively. Rhizomes are harvested by digging the field. Harvesting can be done by hand-picking or carefully lifting the clumps with the help of a spade. The stems are cut about 2-3 cm above the rhizomes. The rhizomes are washed and cleaned thoroughly. The finger rhizomes are separated from the mother rhizomes. The mother rhizomes are stored for seed purpose. The finger rhizomes are cured to extract turmeric. About 20-25 kilogram of fresh turmeric is harvested from 10 m^2 area.

Spinach

Spinach, popularly known as *Palak,* is an annual green with broad leaves. It is rich in vitamin A, vitamin C, potassium, iron and anti-oxidants. Spinach has anti-ageing properties. Spinach prefers a cool climate for growth and development. Optimum temperature range for spinach cultivation is 15 to 30^0 C. Some of the promising varieties are Pusa jyoti, Pusa palak, Pusa harit, Pusa bharati, Punjab green, etc. It can be grown in small beds near the house and pots or container kept on the rooftop or terrace.

Spinach can be grown on varieties of soil having good drainage facility. However, sandy loam and alluvial soils are suitable for growing spinach. It does not perform well in acidic or waterlogged condition. Soil pH should be in the range of 6.0 to 7.0 for obtaining good crop. Land should be thoroughly prepared to get fine soil tilth. Raised beds are prepared at a height of 15 cm with irrigation channels in between. Seeds are sown either by line sowing or by broadcasting method. Line sowing is done at a spacing of 25-30 cm between row to row and 5-10 cm between plant to plant. Soaking of the seeds in water increases the germination.

Well decomposed cow dung or farm yard manure is to be applied @ 25 kilogram per 10 m^2 area at the time of preparation of bed. It requires 60 gram each of nitrogen, phosphorus and potassium for a bed of 10 m^2 area. All the phosphorus and potassium along with half of nitrogen are applied before sowing. Remaining quantity of nitrogen is applied in two equal splits after cutting the leaves. Spinach is very much susceptible to boron deficiency. Application of boron as foliar spray or soil drenching can enhance the yield. However, in backyard cultivation, spinach can be grown organically by using farm yard manure, vermi-compost, vermi-wash, oil cake and other organic fertilizers. One pre-sowing light irrigation is to be given for proper seed germination and better plant growth. During summer months, irrigation is required at 4-6 days' interval, whereas the winter crop may be irrigated at an interval of 10-12 days.

Over irrigation may lead to occurrence of disease and deterioration of quality. Weeds can be controlled to get a clean crop.

Aphid infestation is observed in spinach. It results in stunted growth of the plant and severe reduction in yield. The crop is also affected by the fungal disease *Cercospora* leaf spot. Due to this disease, small circular to semi-circular spots are appeared on the leaves. Downy mildew is another fungal disease affecting this crop. Application of neem oil or neem based pesticide can control the insects and diseases.

The entire plant is usually harvested by cutting just above the ground level. Harvesting is done during cooler part of the day. The crop should be harvested at tender stage before hardening of the stalks. Depending upon variety and season, subsequent cutting should be done at an interval of 20-25 days.

Amaranthus

Amaranthus is an age old green vegetable grown during kharif and summer seasons. It is a warm season crop and can be grown with a temperature range of 18 to 32 ^{0}C. There are various varieties grown in different parts of the country. Some of those are Co 1, Co 2, Co 3, Co 5, etc. It can be taken up in the home garden in small plots or in the pot or container placed on the rooftop.

Amaranthus can be grown on various types of soil having good drainage facility. Well drained loamy soil is suitable for amaranthus cultivation. Land should be thoroughly prepared to get fine soil tilth. Proper levelling is done to obtain uniform seed bed. Raised beds are prepared at a height of 15 cm for sowing of seeds. Seeds are either sown in line or broadcasted. Line sowing is done at a spacing of 12-15 cm between rows. Well decomposed cow dung or FYM may be applied in an area of 10 m^2 during bed preparation. *Azospirillum* and Phosphobacteria may be applied to obtain good yield. Fertilizer can be applied based on soil test results. However, amaranth can be grown organically by using farm yard manure, vermi-compost, vermi-wash, oil cake and other organic fertilizers. One pre-sowing irrigation is to be given for proper seed germination. Irrigation is required at 4-6 days interval during summer season and 10-12 days interval during winter season. Over irrigation may lead to wilting of plant and occurrence of disease & insect. Timely control of weeds facilitates proper plant growth.

Leaf eating caterpillar damages the plant and causes severe yield reduction. The crop is also affected by leaf spot disease. Application of neem oil or neem based pesticide can control these insects and diseases.

Amaranth is usually harvested by cutting just above the ground level. In some cases, the leaves are plucked. About 10 pluckings are done at weekly intervals. Harvesting is done during morning or evening hours to get good quality produce.

Basella (Indian spinach)

Basella or Indian spinach or *poi* grows well in hot and humid climate. It is grown round the year at various parts of the country. It prefers warm climate for growth and development. It can be cultivated in various types of soil. But, it performs well in sandy loam soil with soil pH of 6.5 to 7.0. In home gardens, it can be taken up in any surplus land near the house. It can also be taken up in pots or containers. The sowing time is May-June during kharif season and January-February for summer season crop. Propagation is done by sowing seeds. Sowing is done by broadcasting method or in lines. The field should be well prepared to get fine soil tilth. Addition of 10-15 kilogram farm yard manure or compost per 10 m^2 area enhances yield. Seeds are sown at a spacing of 30 cm X 10 cm for single harvest crop and 60 cm X 30 cm for multi-harvest crop. Provision of staking with the support of bamboo, wooden sticks or wire net at one month after sowing enhances yield. Staking helps in production of superior quality, succulent and attractive leaves. Basella can be grown organically with application of well decomposed FYM, compost, vermi compost, etc. However, application of 20 gram nitrogen, 50 gram phosphorus and 20 gram potash per 10 m^2 area may be done to get good yield. In case of multi-harvesting, about 15-20 gram nitrogen may be top-dressed in 10 m^2 area after every cutting. Weeds are removed manually to get a clean crop. Irrigation should be given at weekly intervals to maintain plant growth. Sprinkling or furrow irrigation is suitable for poi cultivation.

Leaf miners, cutworms and root-knot nematode damage the crop. The damage can be minimized by removal of infested leaves, hand-picking of cut-worms, use of neem/karanja oilcake and soil solarisation. Application of neem oil or neem based pesticide can control these insects and diseases.

First harvest is done by cutting the shoots from 15-20 cm height from ground level. Subsequent cuttings can be done at two weeks' interval.

Drumstick

Drumstick, otherwise known as *moringa*, is a popular plant grown mostly in the backyards. In some areas, it is also grown commercially for earning profit. The leaves, flowers and pods are used as vegetable. The leaves are rich in various vitamins and minerals. Pods are good source of vitamin C, carbohydrate, dietary fibre, protein and minerals. The plant is also used as traditional herbal medicine. It is a warm season crop grown throughout the year. Some of the cultivated varieties are PKM 1, PKM 2, KKM 1, Cylone, Cherry Murungai, etc.

Drumstick grows well in a wide range of soil. Sandy loam soil with pH ranging from 6.5 to 8.0 is ideal for growing drumstick. It is planted during July-October. Drumstick is propagated through seeds, seedlings and stem cuttings. Seed treatment with *Trichoderma viridae* @ 4 g/kg of seed or Thiram @ 3 g/kg of seed can prevent disease incidence. Soil is properly prepared to get rid of stubbles and weeds. Pits of 45 cm x 45 cm x 45 cm size are dug at a spacing of 2.0 - 2.5 m. Fifteen kilogram well decomposed compost or FYM is applied per pit and thoroughly mixed with soil. Generally, planting of two seeds or one seedling or a stem cutting is done in each pit.

Raising of seedling is done in polybags containing pot mixture. About 35-40 days old seedlings can be transplanted in the pits. Stem cuttings of matured plants can also be used as planting material. In home gardens, drumstick is grown by using organic matter like well decomposed farm yard manure, compost, vermi-compost or oil cake. However, in some cases, 70 g nitrogen, 20 g phosphorous and 120 g potassium can be applied as basal fertilizer in each pit. Pinching of the plants at about 75 cm height facilitates production of more branches. Care should be taken for control of weeds by hoeing the field. It is required to apply ash near the plant for better growth. Irrigation is necessary just after sowing the seed or planting the seedling/stem cutting. Subsequent irrigations may be provided at 10-15 days' interval depending upon soil moisture status. Stagnation of water near the plant adversely affects the plant growth. The plant may be pruned at certain interval to get pods at reachable height.

Ratooning is done by cutting the trees at a height of 90 cm from ground level after the harvest of pods. The plants regrow and produce harvestable pods in 4-5 months. Application of 45 g nitrogen, 15 g phosphorous and 30 g potassium along with 25 kg of FYM or compost per pit within a week after ratooning helps in better plant growth.

Drumstick is affected by damping off, leaf spot, powdery mildew, etc. These can be controlled by application of fungicides like propiconazol, mancozeb, copper oxychloride, etc. Hairy caterpillar is the major pest of drumstick. It feeds on the leaves, which results in defoliation of the plant. This can be controlled by foliar spray of neem oil or neem based pesticide. Besides, spinosad or profenophos can be used to control the insect. After settling of caterpillars on the tree trunk, flame can be used to control the insect. A drumstick plant can produce nearly 220 pods per year.

Curry leaf

Curry leaf plays an important role as a condiment in the culinary preparation of various dishes. It is a crop of warm season plant grown in backyard as well as under field condition. The shoot growth may not be up to desired level in places where minimum temperature goes below 13^{0}C. Curry leaf is having many local varieties, which vary from region to region. In home gardens, a few plants can be grown to fulfil the family requirement. This can also be grown in pot or container being placed on the rooftop or terrace.

Curry leaf can be cultivated in various types of soil. However, it grows well in light textured red soil. The field should be prepared repeatedly to get a good soil tilth. Planting is done in the pits at a spacing of 90 to 120 cm. The size of the pit should be 30 cm x 30 cm x 30 cm. Before planting, the pits are filled with 20 kg of well decomposed farm yard manure or compost. Healthy and disease free seedlings are planted in the pits. Normally, curry leaf plants are not provided with chemical fertilizers. The plants are irrigated once in five to seven days depending on soil moisture status. The field should be kept weed free for better plant growth.

Aphids suck the cell sap of the leaves and tender shoots making the leaves unfit for consumption. Sometimes, the plant is affected by leaf spot disease. Foliar spray of neem oil or neem based pesticides can control the insect and disease.

Coriander

Coriander or Cilantro is popularly known as *Dhania*. It is an annual herb, grown for its seeds and green leaves. The seeds and the leaves have a fragrant odour and pleasant aromatic taste due to the essential oil content. The seeds are used as spice whereas the green leaves are consumed raw as salad and chutney. Coriander leaves and seeds are rich source of vitamins and minerals. It has anti-inflammatory properties and acts as an anti-oxidant. It is can be grown throughout the year except hotter months. The popular varieties include Gujarat Corinader-1, Gujarat Coriander-2, Rajendra Swati, CO1, CO2, Rcr-41, Swathi, Sadhana, etc. In urban areas, it can be taken up in small plot adjacent to home or in the container.

Coriander can be cultivated on almost all types of soil with adequate organic matter. Soil with good moisture retention capacity is suitable for this crop. Saline, alkaline and sandy soils are unsuitable for coriander. Land should be prepared thoroughly to get fine soil tilth. Raised beds of 1.2 m width and convenient length with irrigation channels in between are prepared. Coriander is mostly grown during *rabi* season. The seeds are sown during October-November. In some places, late kharif crop can be taken up during August-September. Before sowing, the seeds are soaked in water for 12 to 24 hours to obtain better germination. The seeds are sown in lines at a spacing of 30-40 cm X 10 cm. About 10 kilogram farmyard manure or compost is applied in the bed of 10 m^2 area during land preparation. Coriander is grown organically by using FYM, compost, vermi-compost, oilcake, etc. However, application of 30 gram nitrogen, 30 gram phosphorus and 20 gram potash can be done in 10 m^2 area to get higher yield. Irrigation is given at 10 to 15 days' interval depending upon the soil moisture content. Weeding and hoeing are done 30 days after sowing. If necessary, thinning of the plants can be done to get robust plants.

Coriander is attacked by leaf eating caterpillar, semi-looper and aphid. Powdery mildew is a serious threat to coriander crop. Spraying the crop with *neem* oil or neem based pesticide can manage the diseases and insects. Fully developed and lush green leaves can be plucked for harvesting. Both the leaves and stalks of coriander are used for consumption.

Lettuce

Lettuce is a cool season salad crop. It is one of the few vegetables that can be easily grown in containers or in the kitchen garden. It is usually grown in the winter months in the cooler parts of India. Sowing is done during September-November. The crop performs well in a temperature regime of 13°C - 16°C. The varieties include Great Lakes, Chinese Yellow, Slow Bolt, etc.

Lettuce can be grown successfully in well-drained loose soil rich in organic matter having pH value of 6.0-7.0. It can also be grown in pots using suitable pot mixture. An ideal potting mix is to be prepared by mixing garden soil, sand and vermi-compost or well decomposed farm yard manure. Lettuce is propagated by transplanting of seedlings. Normally, the seeds take 7-10 days for germination. The seedlings are transplanted at 2-3 leaf stage at a spacing of 45 cm x 45 cm. Application of organic manures to lettuce results in good quality produce. Plants are watered as per requirement of the soil. As lettuce is consumed raw, application of chemical pesticides should be avoided as far as possible. Use of neem oil or neem based pesticides can help to control insect and disease infestation.

Lettuce is harvested when the heads are fully developed and of 10-15 centimetre long. Harvesting of lettuce is done during early hours of the morning. The leaves are fresh and crisp during this time. There is emergence of new heads after harvest, which enables to get multiple harvest. The lettuce heads are washed thoroughly to keep the leaves fresh and crispy. It can be wrapped in newspaper and stored in refrigerated condition for longer period.

7

Medicinal Plants

India is known to be rich source of medicinal plants since ancient time. Medicinal plants like *Aloe vera*, *Aswagandha*, basil, mint, cinnamon, etc. are widely used to cure various ailments and for several health benefits. Some of the common diseases like common cold, headache, stomach problem, skin disease, etc. are treated in home by use of various herbs. Medicinal plants can be grown in the backyard or on the rooftop to meet the household demand. Cultivation practices of some of the medicinal plants are described in this chapter.

Tulsi

Tulsi or basil is considered to be a sacred plant with religious importance. It is grown in homes as an aromatic perennial shrub and used for medicinal purpose. Tulsi possess antimicrobial and antiviral properties with capacity to purify the air. It can be grown up to an altitude of 2000 m above mean sea level. This plant can be grown in pots and in home gardens. The leaves and twigs of the plant are used for medicinal purpose. The fragrant leaves and flowers are used in the form of tincture, tea or decoction for curing cold & cough, stomach disorders, skin diseases, etc. The plant is also rich in antioxidant, vitamin and calcium. There are two types of tulsi viz. green type and purple. It can be planted in the pots or in the beds.

Tulsi is a hardy plant and can be grown on any type of soil except highly saline or alkaline soil. Sandy loam soil with good organic matter is considered to be the best for growing this plant. Tulsi cannot tolerate waterlogging condition. The plant can be multiplied through seeds and cuttings. The seeds are directly sown in the soil or seedlings are raised in nursery bed. The nursery should be prepared in partial shade. Seed beds of 4.5 m X 1.0 m X 0.15 m size are prepared by mixing well rotten farm yard manure with finely tilled soil. The seeds are sown during June-July. The seeds germinate in 8-12 days and seedlings are ready for transplanting in about 3 to 4 weeks at 4-5 leaf stage. Tulsi can also be propagated by vegetative method using terminal cuttings. Cuttings of 10-15 cm length with 8-10 nodes are used for vegetative propagation. The cuttings are planted in nursery beds or polythene bags. The rooting process is complete in about 4-6 weeks and the cuttings are ready for transplanting. Transplanting is done in rows at a spacing of 40 cm. About 15 kg of FYM is to be applied in 10 m^2 area at the time of land preparation. If possible, vermi-compost may be applied for better plant growth. Irrigation is provided twice a week till establishment of the plant. Subsequent irrigations may be given depending upon the rainfall and soil moisture status. Weeding is done one month after planting to maintain a clean field. Afterwards, weeding can be done as per the requirement. With increase in canopy cover, weeds are smothered by the plant. However, weeding should be done after each harvest to restrict the weed growth.

Tulsi is not prone to serious pest and disease except some minor pests like leaf rollers. It is also affected by powdery mildew and leaf spot diseases. These can be controlled by spraying neem oil or neem based formulations available in the market. Besides, botanicals prepared from extracts of garlic, *Lantana camera, Clerodendron inerme* and *Calotropis gigantea* are often used for controlling the disease and insect.

Harvesting starts after 90 days of planting. Subsequent harvests are done at every 60-75 days' interval. The crop is harvested at full bloom stage by cutting the plants at 15 cm above ground level. It facilitates regeneration of the canopy to get further yield. On an average, tulsi yields about 10 kg of fresh herbage from 10 m^2 area per annum.

Aloe vera

Aloe (*Aloe vera*) is an important medicinal plant cultivated in India. It is a hardy perennial tropical plant and can be cultivated in drought prone areas. It is widely used in cosmetic industry and preparation of various medicines. A few plants can be taken up in the backyard to get fresh leaves. The leaf extracts are used to cure stomach problem and improvement in skin.

Aloe can be grown in a variety of soils ranging from sandy soil to loamy soil. Well drained laterite to loamy soil is suitable for aloe cultivation. The soil pH ranging from 7.0 to 8.5 is ideal for the plant. Plant growth is adversely affected under waterlogging situation. Aloe has wide adaptability to various climatic conditions. It can be planted in warm humid or dry climate. However, it cannot tolerate extreme cool condition. Some of the important sub-species are *Aloe barbedensis, A. chinensis, A. perfoliata, A. vulgaris, A indica, A. littoralis* and *A. abyssinica.* Varieties namely IC111271, IC111269, IC111280, AL-1, etc. can be grown to get good yield from *Aloe vera.*

It is generally propagated by root suckers or rhizome cuttings. Before planting, the suckers are to be dipped in 0.1 % carbendazim solution for five minutes. Normally, June-July is the ideal time of planting. But under irrigated situation, it can be planted during September-October. The land is to be thoroughly prepared to get good soil tilth. At the time of land preparation, about 15 kg farm yard manure or compost should be applied in 10 m^2 area. Aloe is planted in line at a spacing of 60 cm x 30 cm. About 55 plants can be accommodated in 10 m^2 area. Usually, chemical fertilizer is not applied in aloe. Application of 15 kg farm yard manure or compost in 10 m^2 area every year can help better plant growth. Vermi-compost can also be applied to accelerate the plant growth. Of course, chemical fertilizer can be applied in soils with poor fertility status. Aloe can be cultivated both under irrigated and rainfed situations. Irrigation should be provided immediately after planting. The plant should be irrigated during summer and winter season. Adequate drainage is to be provided during rainy season. At early stage of growth, weeds are to be controlled by manually uprooting or by hoeing.

Sometimes, aloe is affected by fungal disease like leafspot and rotting. These diseases adversely affect the yield and quality of the product. These diseases can be controlled by spraying of neem oil or neem based pesticides. Use of chemical pesticide is restricted in aloe plantation. However, chemical pesticides can be used to protect the crop from disease infestation.

Harvesting is done at 6-7 months after planting. Thick and fleshy leaves are harvested to obtain good yield. Plants are uprooted and processed (gel extraction) within six hours after harvest. Harvesting is done by removing three to four leaves per plant. After harvest, the leaves regenerate and the crop can be maintained up to five years. On an average, 40 kg of leaves can be obtained per annum from an area of 10 m^2.

Rauvolfia serpentina

Rauvolfia serpentina or Indian snakeroot is otherwise known as *Sarpagandha* or *Patala Garuda.* It is a perennial plant growing up to a height of 60-90 cm. It has medicinal properties and the roots are used for curing high blood pressure, stomach problem, asthma, skin disease, etc. This can be grown up to a height of 1200 metre above mean sea level. It grows under temperature regime of 10°C to 40° C and prefers hot and humid condition. *Rauvolfia* grows well in deep loam to clay-loam soil with pH ranging from 5.0 -6.5. The soil should be rich in organic matter and well drained. The soil should be properly prepared for facilitating good development of the roots. Planting two /three plants in the backyard can met the requirement of a household.

Propagation is carried out through seeds, root or stem cuttings. However, propagation through seed gives better root yield. Therefore, propagation of *rauvolfia* is usually carried out by the seed. Around 6-7 g of seed is required for 10 m^2 area. Freshly collected seeds are to be used to get better germination. Soaking the seeds for 24 hours prior to sowing enhances germination and better plant stand. Root cuttings of 5 cm length are planted in nursery beds prepared with mixture of FYM, sand and soil. The beds are kept moist by light irrigation through rose can. The cuttings start sprouting within 20 days after planting and can be transplanted at the desired site. Use of growth hormone in case of propagation through root-cuttings enhances the root formation. The seedlings or cuttings are transplanted at a distance of 45 cm row to row and 30 cm plant to plant. Sowing of seeds or transplanting of root cuttings are done in the month of June-July.

Being a medicinal plant, *rauvolfia* is grown without application of chemical fertilizer and pesticide. Organic manures like farm yard manure, vermi-compost, green manure, etc. are used to meet the nutritional requirement of this plant. *Rauvolfia* can be grown as a rainfed plant. However, irrigation may be given depending on soil moisture condition. This plant can be irrigated at an interval of 15 days during summer months and 25 days during winter months. *Rauvolfia* is a slow growing plant and susceptible to weed competition during initial growth stage. The field should be kept weed free by two to three weedings followed by two hoeings during first year of planting. Flowering and fruiting starts at around 6 months after planting. To obtain good root growth, the flowers are to be removed.

The plant is affected by root knot nematode, caterpillar, white grub, etc. Besides, alternaria leaf spot is a common disease of *rauvolfia*. Bio-pesticides prepared out of neem oil, neem seed kernel extract, neem cake, *dhatura,* etc. are used to control the pest and disease infestation.

Maximum yield can be obtained by harvesting the root at 18 months after planting. The root may go up to 40-50 cm deep in the soil. Harvesting is done by digging up the roots. After digging, the roots are cleaned, washed and cut into 12 to 15 cm pieces for convenience in drying and storage. The dried roots are stored in polythene lined gunny bags in cool dry place.

Aswagandha

Aswagandha or *Asgand* is a medicinal plant widely used in Ayurvedic medicine preparation. It is used to boost immune system, lowers cholesterol, regulates blood sugar and cures bronchitis, rheumatism, inflammation, etc. The root and bark of the plant poses medicinal properties. It is a perennial, bushy and evergreen shrub. Planting two /three plants in the backyard can fulfill the requirement of a household.

Aswagandha grows well in sandy loam soil with good drainage. It can be taken up in slightly alkaline soil with pH of 6.5 to 8.0. The plant performs well in temperature ranging from 20°C to 38°C. Poshita, Rakshita, Jawahar Asgand 20, WSR, Nagori, etc. are some of the commonly grown variety of *Aswagandha.*

The soil is prepared to get a fine tilth, free from clods and derrises. Application of FYM @ 20 -25 kg /10 m^2 area boosts the plant growth. Application of vermicompost and bio-fertilizer results in better plant growth. The plant is propagated by seed or seedlings. About 10 to 12 g of seed is required for 10 m^2 area. The seed rate is around 5 g/10 m^2 area if seedlings are raised in nursery beds and then transplanted in the main field. Six weeks old seedlings are transplanted at a spacing of 60 cm X 30 cm. It is planted during August-September and harvested in May. The crop is mainly grown by use of organic source of plant nutrition. In soils with poor fertility level, low dose of chemical fertilizer may be applied to boost the plant growth and yield. *Aswagandha* plant cannot tolerate waterlogging and needs irrigation at 8-10 days' interval during dry period of the year. Hand weeding is generally done at 4 weeks after sowing/ transplanting to control the weeds. If required, second weeding is done at 3 weeks after first weeding.

Aphid, mite, leaf blight and damping off are some of the major insects and diseases affecting *aswagandha*. Application of *Trichoderma viride* and *Pseudomonas fluorescens* controls damping off disease. Insect and disease attack can be managed by use of neem based pesticides or neem oil.

The crop is ready for harvest in 150-170 days after sowing. The crop attains maturity, when the leaves dry and berries turn orange. The entire plant is uprooted and roots are cut from the stem at 1-2 cm above the crown. The roots are cut into 7-10 cm long pieces to facilitate drying. Seeds are collected from the dried berries by threshing.

Stevia

Stevia is also known as 'honey plant' due to its sweet taste. It is a natural sweetener, which is useful for diabetic patients for balancing insulin in the body. Its leaves are used for preparing various drugs, which are used for treatment of diabetes, tooth cavity repair, etc. It is a perennial herb grown under tropical and sub-tropical conditions.

It is grown in wide range of soils. However, sandy loam to loamy soil with good drainage facility and high organic content is suitable for this plant. The ideal pH range is 6.5-7.5 for better plant growth. The plant grows well in the temperature range of 30-32^0 C. Some of the commonly cultivated varieties are SRB-123, SRB-512, SRB-128, MDS-13 and MDS-14.

Stevia requires well prepared soil with fine tilth for better growth and development. Stevia is transplanted on raised beds. Stem cuttings of 15 cm length are used for propagation. The cuttings are ready for transplanting after 25-30 days of rooting. Stevia seeds can also be sown in indoors for preparation of seedlings. Seedlings are transplanted after 6-8 weeks of sowing in nursery bed. A spacing of 45 cm X 30-45 cm is maintained for transplanting of the cuttings or seedlings. The seedlings can also be transplanted in medium sized pots in home gardens. February-March is the ideal period for transplanting. However, it can be planted in home gardens round the year under controlled condition.

At the time of land preparation, application of FYM vermicompost and *trichoderma* may be mixed with soil to enhance soil fertility. Application of balance dose of fertilizer enhances the yield and quality of leaves. A dose of $N:P_2O_5:K_2O$ @ 27.5:112.5:112.5 g/10 m^2 area is adequate to get good yield. Spraying of boron and manganese is done to obtain good yield. Hand weeding is done to remove weeds from the field. First weeding is done after one month of planting and then subsequent weeding is done every two weeks. The plant requires light irrigation at proper intervals. In summer season, irrigation may be done with an interval of one week. Water stagnation may be avoided in the field during rainy season.

Stevia is affected by various pests and diseases. Sometimes, aphids attack the plant and suck the cell sap resulting in yellowing & premature death of leaves. Diseases like leaf spot and blight affect the plant, which reduces the yield. The insects and diseases can be controlled by spraying of neem oil or neem based pesticides.

Harvesting is done about three months after planting. Subsequent harvesting is done at three months' interval. Stevia is harvested just before flowering when the plants are 40-60 cm in height. At the time of harvesting, 5-8 cm of stem should be left from the ground level to facilitate regeneration. Once planted, stevia can be maintained up to 5 years. The leaves are dried after harvesting. Then the leaves are packed in air tight polybags to increase the self-life.

Centella (*thalkudi/thankuni*)

Centella, commonly known as *thalkudi / thankuni /vallarai,* is quiet easy to grow at home due to low maintenance. It can be grown in beds and pots in balconies or roof tops. It is a perennial aromatic herb that grows well in damp areas. It is used for treatment of skin diseases, stomach disorder, ulcer, etc. The fresh herb has an aromatic odour and bitter taste when crushed. The plant is a popular kitchen herb in some parts of India and used in preparation of chutney, pickle, refreshing drink, etc. It also possesses insecticidal properties.

Centella grows well in moist and marshy lands. It is naturally found around water bodies. Clayey soil with good water holding capacity and high organic matter content is suitable for this plant. It prefers acidic soil and doesn't perform well in alkaline soil condition. Centella prefers partial shade for better growth and quality.

There are three types of centella based on morphological characteristics of the plant. Those are *Centella asiatica*, *Centella cordifolia* and *Centella erecta*. It is propagated by stem cutting with rooted node and few leaves. About 1,00,000 stem cuttings are required to plant one-hectare area. The cuttings can be directly planted in the field or grown in nursery beds for about 4-6 weeks before transplanting. For home gardens, 100 stem cuttings are required to plant in 10 m^2 area with application of well decomposed FYM. Before transplanting, the land is thoroughly prepared to get fine tilth. The field is divided into small size beds for easy operations. About 5 kg of FYM is to be applied/10 m^2 area at the time of land preparation. Light irrigation before planting helps in better establishment of the plants. Subsequent irrigations are given as per the requirement of the plant. Transplanting is done in the month of October at a spacing of 30 cm X 30 cm. After establishment, plant produces off stolon, which proliferates by producing more stolons. Weeds disturb the normal growth pattern of the plant. Hence, weeding should be done at 15-20 days' interval during the initial growth period.

Harvesting is done at fortnightly interval from January onwards. A few stolons are to be left in the field for regeneration. On an average, about 5500 kg of fresh herb and 2000 kg of dry herb can be obtained from one-hectare area. Otherwise, about 6 kg of fresh herb can be obtained from an area of 10 m^2.

Senna

Senna is a small perennial shrub found in wild state in certain coastal parts of India. Senna improves the bowel movement and used for treatment of stomach problem. In urban households, few plants can fulfil the family requirement. Senna is grown in well drained sandy, sandy loam or laterite soils. The ideal soil pH suitable for growing senna is 7.0-8.5. It is a warm weather crop that can be grown under rainfed and irrigated situations. Various varieties are available for cultivation. Some of the suitable varieties are KKM (Se) 1, ALFT-2, Sona, etc.

Land preparation is done before onset of monsoon. A good soil tilth is prepared by ploughing the land 2-3 times followed by cross harrowing and levelling. The field is cleaned by removing crop residues and weeds. The crop is propagated by seeds. About 15 - 20 kg of seeds are required for one hectare area (15-20 g/10m^2). Seeds are soaked in water for 10-12 hours before sowing. Sowing is done during the month of June-July. Sowing can also be done during February-March with assured irrigation facility. Seeds are sown in lines with a spacing of 45 cm X 30 cm.

Usually, medicinal plants are grown without application of chemical fertilizers and pesticides. It is necessary to apply farm yard manure @ 15 tonne per hectare at final land preparation. In the backyards, about 15 kg farm yard manure can be applied in an area of 10 m^2. Application of vermi-compost can also be done for better plant growth. For backyard cultivation, application of $N:P_2O_5:K_2O$ @40:40:40 g/10 m^2 area may be done at the time of sowing. Subsequently, 40 g nitrogen may be top-dressed per 10 m^2 area at 30-40 days after sowing. It is required to remove weeds from the plots manually at early stage of crop growth. During rainy season, irrigation is not required. But, adequate care should be taken for providing drainage.

It is affected by aphid, which causes stunted growth due to sucking of cell sap. It can be controlled by spraying of neem oil or neem based pesticides. Leaf spot is a fungal disease commonly affecting the plant. The plant is also affected by damping off disease, which causes death of the plant at early stage. Application of neem oil or neem based pesticides can control these diseases.

Harvesting is done at two months after sowing and subsequent harvests are done at one-month interval. After harvesting, leaves and pods are dried for 7-10 days to reduce the moisture content to 10 per cent. Under irrigated situation with good management practices, senna can yield about 1500 kilogram of dry leaves and 500 kilogram of pods per hectare. In home gardens, 1.5 kg of dry leaves and 0.5 kg of pods can be obtained from 10 m^2 area.

Brahmi

Brahmi is an important medicinal plant that grows in warm and humid climate. It is commonly found in damp and marshy areas. Bramhi is an annual creeper and grows up to a height of 2-3 feet. All the parts of this plant i.e. seed, root, leaves, rhizomes etc. are used for preparing various medicines to treat anaemia, asthma, diuretic, tumours, snake bite, etc. Besides, brahmi is widely used for controlling cold, cough, fever and diabetes. It is also used for enhancing memory.

This herb is found growing naturally along the canals, water bodies and marshy areas. The plant performs best in alluvial marshy soil. Bramhi prefers acidic soil reaction for good growth. A temperature range of 33^0 to 40^0 C is ideal for optimum vegetative growth of this plant. Pragyashakti and Subodhak are improved varieties of bramhi. Brahmi requires well pulverized and levelled soil. Soil should be properly prepared to get desired soil tilth. The soil is mixed with FYM @ 5 kg/10 m^2 area at the time of final land preparation. The land is irrigated thoroughly before planting. It can be planted in backyard gardens with proper land preparation with addition of well decomposed organic manure.

The plant is propagated by stem cuttings. The cuttings are of 5-6 cm size with few leaves and nodes on it. Those are transplanted in wet soil at a spacing of 20 cm X 20 cm to get maximum yield. Flood irrigation is provided immediately after planting. The ideal time of planting is July-August for better plant establishment.

Inorganic fertilizer dose of N: P_2O_5: K_2O @ 100:60:60 kg/ha is recommended to get good yield. Phosphorus and potash are applied as basal and nitrogen is applied in 3 equal splits at 30, 60 and 90 days after planting. The plants grown in home gardens can be provided with a fertilizer dose of N: P_2O_5: K_2O @ 100:60:60 g/10m^2 to keep the plants healthy. Irrigation is to be done immediately after transplanting for better establishment of the plants. Subsequently, the plants need to be irrigated at 7-8 days' interval, except during rainy season. The field/plot needs to be kept weed free during initial growth phase by hand weeding at every 15-20 days' interval. Later on, weeding may be done occasionally as the plants form a dense mat of vegetation. Grass hoppers are the common pest seen in bramhi, which can be controlled by spraying of neem based insecticides. Brahmi is harvested during October-November, when maximum biomass is produced. About 30 tonnes of fresh leaves and 6 tonnes of dry herbage can be obtained from one-hectare area in a single harvest. In backyard, 30 kg fresh leaves and 6 kg dry herbage are harvested from an area of 10 square meter.

Black Pepper

Black pepper is an important spice crop in the world. It is a perennial climbing vine, which grows up to a height of 10 m. Black pepper has attractive flavour and several health benefits. The fruits of black pepper contain the chemical piperine, which has medicinal properties. A few plants of black pepper can be placed in the home garden to meet the family requirement. It can be planted near big trees, so that the plant can grow with support of the tree. Black pepper is a tropical plant requiring adequate rainfall and humidity. It can be grown up to an altitude of 1500 meters above mean sea level. Pepper can be grown in a wide range of soil such as clay loam, red loam, sandy loam and lateritic soils with a pH value of 4.5 to 6.0. It prefers well drained soil with adequate organic matter.

Several improved varieties of black pepper are cultivated to obtain higher yield. Some of those are Karimunda, Panniyur-1, Panniyur-2, Kottanadan, Narayakkodi, Aimpiriyan, Subhakara, Panchami, Neelamundi, Kuthiravally, Balancotta, Kalluvally, Malligesara, Uddagare, etc. Pepper is propagated by cuttings raised from the runner shoots. Cuttings from the lateral branches are avoided. In case of planting in pots, rooted lateral branches can be used. Cuttings are selected from middle one-third portion of the shoots to get higher yield. Cuttings with 2-3 nodes are planted in nursery beds or polythene bags. Adequate shade is to be provided for better plant growth.

Well drained land and hill slopes are suitable for growing pepper. Pepper is grown as monocrop or in the fruit orchard. It is grown in the plantation of coconut, arecanut, mango, jackfruit, etc.

Pepper cuttings are generally planted during the month of June-July. The land should be ploughed thoroughly to clean the debris and weeds. Pits of size 50 cm X 50 cm X 50 cm are dug at a spacing of 2.5 m X 2.5 m for planting the cuttings. Usually, the pits are prepared about 100-120 cm away from the tree trunk. About 2-3 rooted cuttings are planted in the pit with a distance of 30 cm. Initially, the vines are tied with the stick or pole or tree trunk to enable those for climbing.

Application of 10 kg of well decomposed farm yard manure or compost is to be done in each pit before planting of the cuttings. Application of 100 gram *Azospirillum* per vine is beneficial for plant growth and yield. After three years of planting, 100 g N, 40 g P_2O_5 and 140 g K_2O should be applied per pit per year. These fertilizers are to be applied in two split doses during June-July and September-October. Manuring is done at a distance of 30 cm from the vines. Irrigation may be done at 10 days' interval during winter and summer months. There should not be waterlogging situation during rainy season. Weeding is to be done at early stage of growth. Subsequent weedings are done to keep the field clean. Spraying of NAA @ 40 ppm can be done to increase the size of the berry.

Black pepper is affected by several insects and diseases. Leaf eating caterpillars damage the plant for which the growth is severely affected. Thrips suck the cell sap resulting in stunted growth of the plant. The caterpillars of top shoot borer enter into the young shoots, which results in wilting of the twigs.

Foot rot disease is found in black pepper. Pathogen infects leaves, spikes, collar region and roots. The symptoms are yellowing and sudden collapse of the vines. In case of anthracnose disease, the berries show brown sunken patches during early stages. The infected berries can be distinguished from the hollow berry with characteristic cracks on it. Later on, the berries turn black and dry. Spraying of neem or karanj oil or neem based pesticides can control the spread of these insects and diseases.

Black pepper harvesting is done about 3-4 years after planting. The duration of flowering to ripening stage is about 6-8 months depending upon variety and local climatic condition. Harvesting is done when one or two berries on the spike turn bright or red colour. The spike is harvested by plucking. Berries are separated from the spikes by rubbing with hands or trampling under the feet. Then the berries are dried under sun for 7-10 days till the outer skin becomes black and wrinkled. Full bearing is achieved after 7-8 years of planting. One-hectare plantation can yield about 800-1000 kg black pepper. In home gardens, black pepper can yield about 0.8 to 1.0 kg from 10 m^2 area.

Mint

Mint is an important aromatic cum medicinal plant cultivated across the country. The leaves are widely used for salad purpose as well to add flavour to various foods and drinks. The leaves contain essential oil having 75 – 80% menthol content. The oil is extracted through distillation process. It is used as an ingredient in cough syrups and related pharmaceuticals, cosmetics, mouth washes, etc. Mint is an erect, herbaceous and water loving perennial plant. Some of the commonly cultivated varieties are MAS-1, Hybrid-77, Shivalik, EC-41911, Gomti, Himalaya, Kosi, Saksham, Kushal, etc.

Mint prefers well drained loamy soil with adequate organic matter content. The ideal pH ranges between 6.5 and 7.0 for better plant growth. Sub-tropical areas receiving an annual rainfall of 100-150 cm are suitable for commercial mint cultivation. However, it can be grown in pots in home gardens to meet the daily requirement of the household.

Mint is propagated by stem cuttings and root suckers. Terminal stem cuttings produce more roots. Rooted stem cuttings are planted during June-July in beds at a spacing of 40 cm x 40 cm. The beds are to be provided with farm yard manure or compost for better growth of the plant. Nitrogen, phosphorus and potash may be applied @ 50 gram, 75 gram and 50 gram respectively for 10 m^2 area. At about four months after planting, application of 25 gram nitrogen in 10 m^2 area is useful for better plant growth. Irrigation may be provided based on soil moisture status. About ten irrigations are given during summer season at intervals of 10-12 days. Weeding should be done to keep the plot clean and weed free.

The crop is infested by diseases like stolon rot, fusarium wilt and leaf blight. These can be controlled by application of neem oil or neem based pesticides. Application of neem or karanj oilcake at the time of planting may reduce the risk of disease and insect infestation.

First cutting starts at about 5 months after planting and subsequent cutting at 3 months interval. Under good management conditions, the crop can give good yield for about four years.

8

Plant Propagation

Plant propagation refers to the process of increasing number of plants of a cultivar. Plant propagation is of two types viz. sexual and asexual. Sexual propagation is done by seeds, whereas asexual plant propagation is done by vegetative means such as stem/root cuttings, slips, corm, bulbs, rhizomes, grafting, budding, layering, etc. Good quality seed or planting material of established crop varieties should be procured from authentic source to get desired result. The seed should be uniform in size, free from disease infestation, possess good vigour and having good viability. The quality of the seed is decided based on germination percentage. Seeds can be directly sown to raise a crop or seedlings are raised in nursery beds, which are transplanted later to main plot.

Seedlings can be raised in nursery beds, poly bags or pro-trays. The nursery beds are usually kept one meter wide, convenient length and of 15 cm height. The soil of the bed is prepared by addition of well decomposed bulky organic manure, sand and garden soil in equal proportion. The beds have to be disinfected by exposing the soil to sun (solarisation) and addition of *neem or karanj* oil cake. After sowing the seeds in the bed, those are covered with straw, leaves or compost to protect the seeds from birds and to conserve moisture in the soil. It also helps to maintain the temperature for quick germination. The seeds are sown in lines on the bed. The area required for raising nursery of vegetables and flowers is about $1/100^{th}$ of the main field ($1m^2$ of nursery for 100 m^2 of main plot). The nursery is to be kept clean and healthy by weeding and adequate plant protection measures.

Asexual or vegetative propagation of plants is a method of multiplication from vegetative parts of a mother plant, which reproduces true-to-type plants. In this method, large number of plants can be produced in a shorter period of time. This method is easier, faster and economical. It is the only method of propagation in some plants like rose, tuberose, colocassia, ginger, etc. In case of fruit trees, the time for fruiting is less in case of asexually propagated plants. There are different methods of asexual propagation.

Propagation by cutting

Cutting is the method of using a part of the mother plant for vegetative propagation. It may be stem cuttings, leaf-bud cuttings or root cuttings. The growth promoting substances such as IBA (Indole-3-butyric acid), NAA (Napthalene acetic acid), IAA (Indole acetic acid), etc. are used for inducing rooting of the cuttings.

Stem cuttings: In this case, one-year-old mature stems are collected. Cuttings of 10-45 cm with at least two buds are prepared. A straight cut is given at the base of the shoot just below the node. On the top of the cutting another slanting cut is given 1-2 cm above the topmost bud. This is planted in bed or pot for rooting. Propagation by this method is cheap and easy. This method is adopted in moringa, hibiscus, tagar, rose, etc.

Root cuttings: Root cuttings of the plants are used for propagation. One cm thick and 10-15 cm long roots are used for the purpose. This method is adopted for propagation of pointed gourd, blackberry, etc.

Leaf-bud cutting: A piece of leaf along with a bud is used for propagation. Rhododendron, camellia, rubber plant, dracaena, etc. can be propagated by this method.

Propagation by Budding

Budding is a technique of inserting single bud from the desired scion in the rootstock to get a new plant. Budding techniques are of different types such as T budding, inverted T budding, patch budding, ring budding, etc. Budding is used for propagation of rose, hibiscus, bougainvillea, etc. The scion and rootstock should be compatible for success.

T-Budding: T-budding is most commonly used for rose, apple, pear, hibiscus, etc. One T shape cut is given on the rootstock. The bud along with a small portion of bark is taken from the desired plant. The bud is carefully inserted in the bark cut and wrapped tightly with polythene strip. After successful union and emergence of twigs from the bud, the unwanted branches of the plant are removed to enable better growth of the new branch.

Inverted T-Budding: In case of inverted T-budding, the cut is given on the rootstock in the shape of an inverted T. This is suitable in areas receiving high rainfall or for the species with more oozing of cell sap after receiving the cut. It helps in better drainage of fluids and healing of the cut area. Other techniques are same as T budding. This is used for chest nut plant.

Patch budding: A rectangular patch of bark is removed from the rootstock and a patch of bark of same size containing bud from desired plant is placed on it. This is suitable for the trees having thick bark such as mango, jack fruit, jamun, aonla, etc.

Ring budding: This is also known as annular budding. In this type of budding, a complete ring of bark is removed from the rootstock. A similar ring of bark containing a bud is removed from the desired plant and is inserted on to the rootstock. It is useful for the plants like ber, peach, mulberry, etc.

Chip budding: Chip budding is suitable for mature buds. It is used for the species, where bark does not slip easily. Plants like grape, apple, pear, etc. are propagated by this method.

Propagation by Grafting

It is the process of uniting two separate plant parts so as to prepare one desired plant. In this case, rootstock and scion are used to create a new plant. Rootstock is the mother plant grown, which acts as the base of the new plant. It is prepared from seeds of any variety of the plant. Scion is the small portion of the desired plant, which is inserted in the rootstock to get the new plant. The scion is inserted or attached with the rootstock to create a unique plant with improved characters. The typical characteristics of the scion are expressed in the new plant. The best quality scion wood comes from the shoots of previous season. Scions should be cut with sharp knives and placed immediately in moistened cotton bag. To get best results, the scion wood should be prepared in the same day for grafting. If scion is stored longer, its vitality reduces. Healthy scions free from insect, disease or injury are to be selected.

There are various types of grafting such as cleft grafting, bark grafting, side-veneer grafting, splice/ whip grafting, saddle grafting, bridge grafting, inarching, etc.

Cleft grafting: Cleft grafting is one of the simplest and most popular forms of grafting This type of grafting is done, when both scion and rootstock are not in active growth stage. Cleft grafting may be performed on main stem or on lateral branches. The diameter of rootstock used for cleft grafting is more than that of scion. The ideal girth of rootstock is 5-7 cm and it is at 45 cm above ground level. Scions of 15-20 cm long collected from three-month old terminal shoots are suitable. The stock should be given a 5 cm deep perpendicular cut with a clean clefting knife and a mallet. The scion is inserted in the cut immediately after removal of knife. Then the cut surface is to be sealed with grafting wax to prevent drying. It is suitable for apple, cherry, pear, etc.

Bark grafting: Bark grafting is used for rootstock of larger diameter (10 to 30 cm). It is practised during the period of the year, when the bark can be easily detached from the wood. The rootstock is cut with a sharp saw, leaving a clean cut. A vertical slit of 5 cm long is made through the bark for insertion of scion. Several scions are prepared with tapered wedge on one side. The bark is slightly loosened and the tapered surface of the scion is inserted under the bark. The scion is inserted at every 8-10 cm distance around the bark of rootstock. The exposed surface is sealed with grafting wax to prevent entry of moisture and pathogens. Normally, bark grafts form weak unions and require staking at the initial years of establishment.

Side-veneer grafting: In case of side-veneer grafting, the rootstock is grown in pots one season before grafting and allowed to stay in dormant stage. A shallow downward cut of about one inch is given at the base of the potted rootstock to expose a flap of bark with some wood still attached. The scion chosen should be of same diameter or slightly smaller than the rootstock. A slanting cut of about one inch is made at the base of the scion. The cut surface of the scion is inserted against the cut surface of the rootstock. The scion is held in place using a rubber grafting strip, tape or grafting twine and the graft area is to be sealed with warm grafting wax. The rubber or twine should be removed after the union is established.

Splice/whip grafting: Splice grafting is used to join a scion with a rootstock or with an intact root piece. This is usually practised in herbaceous plants with a stem diameter of 1.5 cm that forms tissues easily. In splice grafting, both the stock and scion should be of the same diameter. Diagonal cuts of 2-2.5 cm are made both on rootstock and scion. The scion is fitted to the stock by wrapping the joint securely with a rubber, grafting strip or twine. The grafted area is sealed with grafting wax. Watering is done sparingly until the graft unites. Removal of the wrapping material should be done as soon as the graft heals.

Saddle grafting: In saddle grafting, the stock may be either field-grown or potted. Both rootstock and scion should be of equal thickness with the diameter within one inch. The top of the rootstock should be cut in a manner to resemble an inverted V. The length of the cut is to be 2-3 cm. The scion should be given a V shaped cut and inserted in the rootstock so that maximum of tissues will make contact. Then the graft is wrapped with a grafting twine, tape or strip and sealed with grafting wax.

Bridge grafting: Bridge grafting is done to repair the diseased or damaged portion of a plant, usually near the base of the trunk. The bridge graft provides support and facilitates water and nutrients to move across the damaged area. Bridge grafts are usually done just before active plant growth starts. The scions should be straight and about twice the length of the damaged area. A tapered cut of 4-5 cm is made at each end of the scion. The damaged tissues are to be removed and grafting is done on healthy stem. Grafting is done by cutting the bark on the rootstock below the damaged portion and inserting the scion. Bridge grafts should be spaced about 8-10 cm apart across the damaged area. All the grafts are sealed with grafting wax or grafting paint. Any bud or shoot emerging on the scion during and after the healing period are to be removed.

Inarching: Inarching is used to support a damaged portion of the stem. The stock is brought close to the scion. A slice of bark of about 6-8 am at a height of 20 cm above ground level is removed from the rootstock. A similar cut is made on the scion and both are joined together and tied with polythene strip. It is used in guava, mango and other plants. It is useful to rejuvenate old and damaged trees and impart a new look.

Propagation by Layering

It is a method of vegetative propagation, where roots are induced to develop on plant parts while those are still attached to the parent plant. Layering is of different types such as ground layering, tip layering, serpentine layering, air layering, mound layering, trench layering, etc.

Ground/simple layering: Plants with low growing and trailing stems can be propagated through ground layering. A healthy, young stem close to the ground is bent down and a small cut is given in the portion of the stem to be layered. The leaves are removed from area and it is buried in the soil. The tip of the stem is kept in open and allowed to grow. After emergence of roots,the layered portion of the plant along with the tip is detached from the mother plant and used as planting material. Lemon and cashew nut can be propagated by this method.

Tip layering: Tip layering is quite similar to simple layering. In this case, the tip of the plant is buried in soil at a depth of 8-10 cm. After some days, the buried tips emerge out of the soil and grow upward with roots in the buried portion. This regenerated portion can be detached from mother plant and used as planting material. Plants propagated by tip layering are black berries, raspberry and dewberry.

Air layering: Air layering is also called as goottee. In case of air layering, stems are layered above ground. This method is useful for plants, which are difficult to bend down to the ground. The leaves are removed from a healthy stem below the growing tip and the bark is removed. It is covered with rooting medium such as moss and wrapped with transparent polythene strip of 200-300 gauge. The plant produces roots on the layered stem, while it is still attached to the parent plant. After about six to eight weeks, there is formation of roots. The portion of the stem below the roots are cut off and used as planting material. This method is suitable for litchi, guava, jackfruit, sapota, cashew nut, rubber plant, etc.

***Compound (serpentine) layering*:** It is a modified form of simple layering, where the stem is alternatively covered in the soil and exposed to get multiple plants from a single stem. The buried portions of the stem are to be given a cut for rooting. Each part of the exposed portion of the stem should have at least one bud to develop new shoot. This method works well for grapes.

Mound /stool layering: Mound / stool layering is suitable for thick stemmed and closely branched trees like apple, pear, etc. In this case, plants are cut at 2.5 cm above the ground level during the dormant season. New shoots will emerge within two months. The new shoots are mounded with soil (15-20 cm height) as those grow. Roots will develop at the base of these young shoots within 20-30 days. The rooted stools are separated from the mother plant after two months for planting.

Trench layering: Trench layering is the process of growing a plant or a branch in a horizontal trench. The branch is buried in the trench and covered with soil. New shoots along with roots develop from the buried portion, which is detached and used as planting material. This method is used for propagation of apple and walnut.

Other Methods of Propagation

Plant parts like bulb, corm, tuber, rhizome, pseudo-bulb, etc. are also used for propagation of various plants.

Bulb: It is a modified shoot used for propagation of certain plants like lily, tulip, tuberose, etc. A bulb has a short disc-like stem with a growing point. The stem is covered with several layers of fleshy leaves with stored food material.

Corm: Corm is the modified underground stem used for propagation of plants. The corms are cut into individual sections having at least a single bud. Each section of the corm generates a new plant. Plants like colocasia, gladiolus, etc. are propagated by corm.

Tuber: It is a modified underground fleshy and enlarged stem with eyes on it. The eyes contain buds, which give rise to new plants. The entire tuber or cut pieces with one or two eyes are used for propagation. Plants propagated by tubers include dahlia, caladium, potato, etc.

Rhizomes: This is a modified underground stem which grows horizontally below the soil surface. The stem is fleshy and compressed. Plants like ginger, turmeric, etc. are propagated by rhizomes.

Tools used for Plant Propagation

Budding knives: Special knives are used for budding and grafting of plants. It is a small knife made of thin metal and used for making incision on the rootstock for insertion of the graft. It is also used for removing the bark and preparation of buds.

Secateurs: Secateurs are gardening tool and look like a pair of long, heavy scissors. It is used for cutting the branches and stems of plants.

Budding rubber**:** It is a strip of pliable rubber of 10 to 20 cm long used during budding.

Grafting strip: It is a rubber strip used to hold scions during grafting process. Grafting strips are thicker and less flexible than budding rubber.

Grafting twine: It is used for wrapping of the graft joints and cambia for success of grafting. It is made up of jute or raffia.

9

Preservation of Fruits and Vegetables

Fruits and vegetables grown in kitchen garden or in the field can be preserved in various forms like juice, squash, puree, pickle, jam, jelly, *murabba*, paste, ketchup, sauce, etc. for year round consumption. Seasonal produce of vegetables and fruits can be preserved through various methods for consumption at a later time.

Juice

Fruits like pineapple, mango, orange, lemon, etc. can be used to prepare juice, squash and other ready to serve beverages. Beverages can be prepared by following various steps. The fruits are washed thoroughly and pat dried with clean cloths. Then the peels are removed wherever necessary. Peeling can be done by hand or by knife. The juice is extracted by using juice extractor or by netting. The extracted juice is then strained using a muslin cloth or mosquito net. To enhance the taste and shelflife, the juice is mixed with sugar syrup. The sugar syrup is prepared by dissolving sugar in water and boiling it. After the syrup is ready, it is allowed to cool and strained through muslin cloth. Then the extracted juice is added with the syrup and citric acid. Preservatives like potassium meta-bisulphite or sodium benzoate is added for storage of the juice. If necessary, essence can be added @ 2 ml/litre of juice. The juice is to be stored in a narrow mouth bottle by leaving a head space of one to two centimetres. The cap of the bottle is to be sealed with wax.

Jam

Jam constitutes an important segment of fruit products. It is prepared from fruit pulp added with sugar. It is consumed along with bread, *chapati* and other food items. The consistency of jam depends upon the type of pectin present in the fruits. Fresh or frozen fruits can be used for preparation of jam. Matured, well ripen and healthy fruits are suitable for this purpose. A good jam should be palatable and appeal to the eye. Over ripen fruits are not suitable for preparation of jam. The fruits are thoroughly washed with clean water to remove all the dirt and dusts. It helps to clean the residues of pesticide and ripening chemicals off the fruits. Chlorinated water can also be used for washing the fruits. The skin of the fruit is removed by hand peeling or with the help of a sharp knife. Then the fruits are cut in to small pieces and boiled in water for extraction of pulp. About 250 ml of water is used per kg of pulp and boiled for 45 minutes. The pulp is then strained with the help of muslin cloth or mosquito net. Sugar is added to the strained pulp at 1:1 pulp: sugar ratio and the mixture is boiled to a reasonably thick consistency till there is shiny appearance of the mixture. The right consistency is judged, when the mixture doesn't drop down while picking up with a stick or spatula. Then the mixture is cooled down and citric acid is added @ 4g/kg of pulp. Wide mouth bottles are used to preserve the jam. The bottles are filled up to the rim and allowed to set. Then the cap of the bottle is sealed with wax. Jam can be prepared from fruits like apple, guava, papaya, mango, pineapple, *sapota*, plums, banana, pears, etc.

Jelly

Jelly is the viscous or semisolid processed food prepared from fruits. The mixture of fruit pulp and sugar should be in the ratio of 9:11 by weight. The mixture of pulp and sugar is boiled till a transparent and well set concentrate is obtained. The finished product should be soft enough and retain the original flavour of the fruits used. Preparation of fruit jelly is an easy process without requiring special equipment.

The essential ingredients required for preparation of fruit jelly are matured, well ripen & healthy fruits, pectin, sugar, acid, preservatives and water. There should be certain level of acidity in the juice for formation of jelly. If the fruit juice is not acidic, the gel like consistency will not be obtained. On the other hand, too much acidity will adversely affect the quality of jelly. If the juice is not acidic enough, one tablespoon of lemon juice may be added to each cup of fruit juice. Addition of sugar helps in gel formation, contributes flavour to the jelly and serves as a preservative.

Generally, jelly is prepared by standard or long boil method. This method should be used only for fruits that contain adequate quantity of pectin. In this case, the fruit juice and sugar are boiled for long time to form a gel. This type of jelly has rich flavor.

The fruits are washed thoroughly and cut into small pieces before cooking. Peeling of the skin helps to get better quality jelly. Some fruits require addition of water during cooking. The fruit pieces are boiled until getting soft. Constant stirring is to be done to prevent sticking to the bottom of the dish. When the fruit becomes soft, it is strained with the help of a muslin cloth to get a smooth pulp. Then sugar is added in the pulp and the mixture is boiled again.

The mixture is stirred until the sugar is dissolved. It is boiled till the desired consistency of jelly is obtained. The consistency is adjudged by the following process. A cool metal spoon is dipped into the boiling jelly mixture and taken out of the dish. It is held at about 30 cm above the pot. The correct consistency of the jelly is obtained when two large drops slide together and form a sheet at the edge of the spoon. Then it is cooled down and poured into a wide mouth glass/plastic container. The cap is tightened after complete cooling and it is sealed with wax.

Marmalade

Marmalade is prepared from orange, lemon, pomelo, grapes, etc. For preparation of marmalade; matured, well ripen and disease & blemishes free fruits are taken. Over ripen fruits are not suitable for marmalade preparation. The fruits are washed thoroughly with clean water. The bright outer covering of orange or lemon is taken out with the help of a vegetable peeler or sharp knife. The peeling is to be done carefully so that only the outer covering is removed without the inner white pith underneath the outer cover. If the pith is peeled along with the zest the taste of the marmalade becomes bitter. The orange is cut and the thick white pith is removed. Then the juice is to be extracted from the fruits. The bright outer cover of the fruit peeled earlier has to be cut into thin slices. The juice, sugar, water and the slices of the orange or lemon peel is to be put in a heavy bottomed pan and boiled. For half kilogram orange, equal quantity of sugar and 1.25 litre of water are to be taken. The mixture needs to be boiled for about one and half to two hours. A drop of the boiling mixture is put on a chilled plate to decide whether the marmalade is ready. The right consistency is reached when the mixture doesn't spread out. The cloth containing the membrane and seed is to be removed and the hot marmalade is poured in clean glass containers leaving half inch of headspace at the top of each jar. The jar is covered with lid and kept in refrigerator. Generally, the marmalade contains more sugar than jam and jelly.

Squash

Squash is prepared by mixing fruit juice, water and sugar in definite proportion. Sometimes, artificial food colour and flavour are added to enhance the look and market value of squash. Fruits like mango, apricot, lemon, lime, litchi, orange, pineapple, jamun, watermelon, etc. are used for preparation of squash. Squash is prepared by extracting the juice or pulp of the fruits. The process of extraction of juice/pulp varies depending on the fruit. The juice extraction processes of some of the common fruits are stated here.

Lime and lemon: Fresh, fully matured and healthy fruits are taken for juice extraction. Those are rinsed thoroughly with clean water. Then the fruits are cut into halves with a sharp stainless steel knife. The juice is extracted with a juice squeezer and strained through muslin cloth/plastic strainer to remove seeds.

Orange: Oranges are peeled to separate the carpels/segments. The segments are then put in screw type juice extractor. The juice is collected in a clean pot and strained through muslin cloth/ strainer.

Mango: Fresh and ripe mangoes are selected to prepare squash. Those are rinsed thoroughly with clean water and pat dried before extraction of pulp. The pulp is extracted by pressing with hand or by passing through a pulper.

Litchi: Healthy and ripe litchi fruits are taken for preparation of squash. The seeds are removed and juice is extracted by passing through juice extractor or pulper. Juice can also be extracted with the help of mixer grinder used in kitchen. The juice is then strained through muslin cloth/ plastic strainer.

Plum/apricot/jamun: Well matured shiny fruits of plum or apricot or jamun are selected to prepare squash. After washing the fruits, those are boiled with little water. The boiled fruits are allowed to cool down and then passed through pulper to extract the pulp and separate the seeds.

Watermelon: Ripe fruits are used for squash preparation. The peels are removed and seeds are separated. The pulp is passed through pulper or mixer grinder to extract the juice.

Pineapple: The crown and the peels are removed from the fruits with the help of a sharp knife. The pith of the pineapple is to be discarded. Then the fruit is cut into small pieces. The small pieces of pineapple are put in a mixer grinder or food processor to get the pulp. The pulp is strained through coarse muslin cloth or strainer to extract the juice.

After extraction, the juice is mixed with sugar syrup. Sugar syrup is prepared by mixing sugar and citric acid with water. In general, 1.8 kg of sugar is mixed with 1.2 litre of water and 25-30 g of citric acid for preparation of sugar syrup for one litre of fruit juice/pulp. The mixture is boiled, strained and allowed to cool. The fruit pulp or juice is mixed with sugar syrup. Thereafter, about 2 gram of potassium meta bi-sulphate (KMS) or 4 gram of sodium benzoate is added to the mixture for long term preservation of the squash. If required, permitted edible food colour and flavour may be mixed to enhance the look and flavour of the product. Pre-sterilized glass or plastic bottles are used to store the squash. The bottles are filled and sealed properly.

Pickles

Pickles are most common method of preserving fruits and vegetables. Fruits like mango, lemon, aonla, ber, jackfruit, tamarind, etc. are preserved as pickles. Besides, various seasonal vegetables like radish, carrot, tomato, cauliflower, chili, drumstick, etc. are preserved by preparation of pickles for year round use. Well matured fruits at green stage are suitable for pickle preparation. The fruits are to be washed with clean water to remove dust and dirt particle. Those are to be pat dried for removing excess water adhered to the fruit. The fruits like jackfruit are peeled to remove the skin. Ber and cut pieces of dilenia are put in brine solution for about one to two weeks. The fruits and vegetables are mixed with salt and sun dried for few days to reduce the water content. However, in case of instant pickles, the fresh vegetables/ fruits are marinated with spices, oil, vinegar, etc. and allowed to absorb the spices for two to three days.

Pickles can be prepared in various processes depending on the types of fruit or vegetable used and as per personal taste. It can be an instant process or a long term processing by repeated sun drying and mixing of oil and spices. The pickles prepared by thorough processing can be stored for a longer period. This type of pickle retains its flavour and taste for two to three years, though the colour becomes slightly darker after few months. Preparation of instant pickle is quick and easy. The shelflife of instant pickle is shorter.

The common spices used in preparation of pickles are salt, asafoetida, turmeric, red chili, mustard, fenugreek, kalonji or black cumin, fennel seeds, carom seeds, etc. Besides, other spices can also be added as per taste. The spices are powdered and added to the processed fruits or vegetables along with oil, vinegar, acetic or citric acid to enhance the taste and shelflife of pickle. The oil used for pickle is to be heated till it smokes and then allowed to cool down before adding it to the pickle. For sweet pickle, sugar or jaggery is added along with the spices to get desired taste. Wide mouth glass bottle or jar is used to store pickles.

Petha

Petha is a popular sweet, which is prepared from ash gourd. Fully ripe gourds with tough texture are used for preparation of *petha*. The gourds are washed with clean water and the peel is removed with the help of a sharp knife. Then it is cut into small slices and the seeds are removed. The inner soft pulp is used for *petha* preparation. The peeled slices are cut into desired shape and size. The cubes or pieces of gourd are pricked with a stainless steel fork or wooden pricker. Those are completely immersed in fresh lime water. The lime water is prepared by mixing 60 gram of quick lime per litre of water. The mixture is stirred vigorously and allowed to settle down. The upper part of the liquid is decanted and filtered. The pricked cubes or pieces of gourd are soaked in the lime water for 3-4 hours depending upon the softness of the pulp. Soft pulps are to be soaked for longer period as compared to tough textured pulp. The lime water is drained off and the fruit pieces are washed thoroughly in running water.

Then the pieces of gourd are boiled in water for about 15-30 minutes to enable the gourd pieces for absorbing the sugar syrup. After boiling, the water is drained and the pieces are spread on a clean white sheet for removing moisture. Sugar syrup is prepared by dissolving two parts of sugar in three parts of water. The solution is boiled and the impurities are removed by a ladle. Then the solution is filtered through a thick cloth to obtain a clear syrup. Then the syrup is boiled again till it becomes thick enough. The boiled and dried fruit pieces are to be added to the boiling syrup at the ratio of 1:3 (fruit: syrup). The mixture is cooked for some more time. The entire preparation is allowed to remain as such for overnight. Then excess syrup is drained off through a stainless steel sieve. The pieces of syrup coated gourd are dried in shade and rolled on sugar powder. Those are the finished sweet called petha and stored in clean container.

Tomato Products

Tomato can be processed at home to prepare puree, paste, ketchup and sauce; which can be preserved for use round the year. Fully ripen, deep red coloured and firm fruits are used to prepare processed products. The fruits should be free of disease, pest and blemishes.

***Tomato puree*:** Tomato puree is the concentrated tomato juice or pulp without skin or seed. The puree may be salted or without added salt. Tomato juice or pulp is to be concentrated to get 9.0 to 12.0 per cent total solids for preparation of tomato puree. Tomato juice or pulp is extracted and filtered to remove the skin, seeds and large coarse pieces to get uniform juice or pulp. The juice or pulp is then boiled in a wide mouth heavy bottom container to evaporate water till the volume is reduced to half of the juice/pulp taken. Alternatively, the end point is determined by a hand refractometer by measuring the total soluble solids. The puree is then filled into cans and kept in boiling water for 20 minutes. The processed cans are cooled immediately by dipping in cold water. The cans are then stored in dry and cool place.

***Tomato paste*:** Tomato paste is the concentrated tomato juice or pulp without skin and seed. Unlike tomato puree, the paste contains more than 25 per cent of tomato solids. If the tomato solid level is of 33 per cent or more, then it is called as concentrated tomato paste. For preparation of tomato paste, tomato juice or pulp is first heated in heavy bottomed pan to reduce the moisture level and to keep the total solid levels in the range of 14-15 per cent. Subsequently, concentration is increased further by cooking in open kettle. Common salt or basil leaf may be added to prevent excessive foaming, burning and sticking of the pulp. The pulp is stirred till bright red coloured thick paste is formed. It is cooled and stored in clean glass jars.

***Tomato ketchup and sauce*:** Ketchup and sauce are the most popular and common products to preserve tomato during bumper production. The ketchup and sauce are prepared by blending concentrated tomato juice with sugar, salt, vinegar, spices, condiments, etc. Selection of good quality tomato is important for preparation of ketchup as it affects the quality and shelf-life of the finished product. Fully ripen deep red coloured tomatoes, free from disease, pest & blemishes are suitable to obtain good quality product. Green and yellow coloured portions of the fruit should be discarded.

The fruits are thoroughly washed and pat dried with a clean cloth. Then those are cut into small pieces and put on flame to boil in a heavy bottomed pan. After through boiling, the pulp is strained to get a uniform puree, free of seeds and skin. The thin watery puree obtained by the process is further boiled to get desired concentration. The spices like red chili powder, black pepper, nutmeg, clove, cinnamon, cardamom, mace and cumin are added as per taste. Besides, onion, ginger and garlic are also used in preparation of ketchup. The spices are coarsely ground/ chopped and tied loosely in a muslin cloth and the packet is placed in the tomato juice during boiling. It has to be pressed intermittently to release the flavour in the sauce.

Sugar is to be added to improve the taste of the product. About one-third of sugar is added to the strained pulp at initial stage of boiling. This retains the natural colour of tomato. Rest of the sugar is added a minute before reaching final concentration. If all the sugar are added in the beginning, the colour of the sauce may turn brown due to caramelizing of sugar. Little quantity of salt is also added at the end of boiling process of the ketchup. Glacial acetic acid is added to improve the flavour and extend shelflife of the sauce/ketchup. Sodium benzoate is the most common preservative used in tomato ketchup/sauce. After attaining the desired consistency, the ketchup is put in the bottle. It prevents browning of the sauce and loss of vitamin during subsequent storage. Bottled products are stored under cool temperature in dry place.

Ginger Products

***Dry ginger*:** Ginger can be stored for long period in the form of dry ginger. Healthy, matured and partially dried rhizomes with hard skin are used to prepare dry ginger. The rhizomes are thoroughly washed in clean running water to remove soil and dirt particle. Light peeling of the rhizome is done within 12 hours of harvesting without rupturing of oleoresin cells below the skin. The peeled rhizomes are cut into small pieces of 50-60 grams. The cut pieces are again washed in clean water and dipped in 20 per cent lime slurry for at least 4-6 hours. Then those are to be dried in a clean surface under the sun. The process of drying and dipping in lime slurry is continued till uniform white colour is obtained. It takes about 10-15 days till the moisture content of dry ginger is 8-10% level. The dried ginger can be polished by rubbing with gunny bags to get smooth product. It is to be packed in air tight containers and can be stored for 6-8 months.

***Salted ginger*:** Ginger can be preserved as salted slices. Tender rhizome with soft skin is preferred for preparation of salted ginger. The rhizomes are cleaned with water to remove the dirt, soil & roots. The skin is peeled with stainless steel knife. After peeling, the rhizomes are washed again under running water and dried with a cloth. Then the rhizomes are cut into thin slices and dipped in 16% salt solution for few hours. The salted slices are sundried for 2-3 days and packed in airtight containers. Dried ginger can be used as a spice and also eaten to cure digestive disorders.

***Ginger paste*:** Matured rhizomes are taken for preparation of ginger paste. The rhizomes are washed thoroughly under running water to remove the soil, dirt particles and roots. The skin is peeled without damaging the oleoresin cells underneath. The peeled rhizomes are cut into small pieces and washed again. Then the sliced ginger pieces are put in filtered water and boiled for 5-7 minutes. After the slices are blanched, the water is drained out. After cooling, the slices are mixed with common salt @ 2 gram per 100 gram of rhizome and grinded to fine paste in a mixer grinder. Sodium benzoate is mixed @ 0.1 gram per 100 gram of ginger paste to enhance the storage period. The paste is to be stored in sterilized glass bottles.

10

Miscellaneous Activities

Potted Plants

In urban areas, scarcity of space compels the dwellers to take up gardening on rooftop, terrace, balcony or small space adjacent to the house. Usually, plants are grown in pots or containers for rooftop gardening. Big size pots with diameter of 30 cm or more are required for growing rose plant. It is necessary to prefer tall pots to obtain appropriate growth of the plant having deeper root system. Now-a-days, pots of different materials such as clay, cement and plastic are used for growing potted plants. Usually, clay pots are better than plastic pots as clay is slow in transfer of heat thereby maintaining the root zone cooler. Of course, plastic pots are lighter and suitable for use on the rooftop as those put less load on the building structure. Light coloured plastic is preferred to dark colour as those generate less heat. Although heavier, cemented pots are more durable than earthen or plastic pots. The pots should have adequate drainage facility with holes in the bottom. Besides, a layer of gravel or medium-size stone of one-inch thickness are to be placed at the bottom of the container to facilitate drainage.

In urban gardening, pot or container plays an important role to grow plants. The pots or containers may be made up of clay, cement or plastic. The pots and containers should be large enough to accommodate the roots of plant. The pot should be able to hold the soil and moisture required for growing plants with leaf, flower and fruit. The pot should bear the weight of above ground part of the plant and support the height of the plant.

In case of small plants, more than one plant can be grown in a container. Small pots can also be used to accommodate small plants. Soil depth of at least 20 cm or 50 cm can hold sufficient soil and moisture to support a standard vegetable plant. Smaller containers holding less soil may require more frequent watering.

Proper drainage facility is essential for growing plants in containers. Containers with less than 25 cm diameter should have a hole of about 1.2 cm diameter to with more facilitate good drainage. Containers with more than 25 cm diameter need about two to four holes. Size of container varies depending on type of plant.

Summer season is a tough time to manage the plants due to high temperature. Of course, some plants like succulents and cacti are well adapted to high temperatures. Usually most of the plants benefit from sunlight, but the rising heat in the cities is more harmful for the terrace gardens. In case of potted plants, the crisis is more acute than the garden plants.

Potted plants require additional care to maintain the plant growth. The plants in plastic pots are more susceptible to high temperature than the plants in earthen pots. It is important to water the plants properly to minimize the effect of heat. However, over-watering is to be avoided to get healthy plants. If possible, high humidity can be maintained in the immediate environment through misting or spraying of water. Keeping the pots on the pebbles and watering the pebbles frequently may make the micro-climate humid. If possible, the plants may be moved to shade under the tree to get rid of scorching sunlight. Alternatively, shades may be created by polythene sheet or thatching materials.

The pots are to be covered with gunny bag, old cloth or paper to protect the pots from heat effect. Repotting should be avoided during summer season as it may be harmful. As repotting is associated with trimming of some roots and leaves, it will result in additional shock for the plant. Pruning and training activities are to be avoided during summer season. Application of adequate quantity of organic matter in the soil helps the plants to tolerate more heat.

Nutrient application

Fertilizing potted plants is very important in order to maintain plant growth and development. Fertilizing the container plants is even more important than those grown in a garden. Since the plant roots cannot go deeper or spread outside the pot, it is required to provide adequate quantity of nutrient to meet the requirement of the plant. Potted plants completely depend on outside supply to provide all the nutrients to grow and thrive. Hence, it is important to fertilize these plants on a regular basis throughout the growing season.

Plant nutrients play a vital role in growth and development of any plant. Plants require 17 essential elements for maintaining growth and production. These elements are divided into various groups such as primary, secondary and micro nutrients. The primary plant nutrients are Carbon (C), Hydrogen (H), Oxygen (O), Nitrogen (N), Phosphorous (P) & Potassium (K); the secondary nutrients are Calcium (Ca), Magnesium (Mg) & Sulphur (S) and the micronutrients are Iron (Fe), Manganese (Mn), Zinc (Zn), Copper (Cu), Boron (B), Molybdenum (Mo), Chlorine (Cl) & Nickel (Ni). The plants obtain carbon, hydrogen and oxygen from water and air. Other nutrients are received from the soil. It is necessary to apply the nutrients in the form of organic matter and chemical fertilizer to boost plant growth. Adequate care should be taken to provide nutrients to the potted plants. It is preferred to apply plant nutrients in the form of organic matter such as well decomposed farm yard manure, compost, leaf mould, vermi-compost, vermi-wash, oilcake, etc. Chemical fertilizer can be applied in the form of water soluble fertilizer.

Liquid fertilizers can also be used in potted plants. Liquid fertilizer products are absorbed into plants through a liquid solution more rapidly. Water soluble fertilizers provide required nutrients for improving growth and vigour of the plant. However, application of organic or natural based nutrients can reduce the risk of fertilizer scorching and provides balanced nutrition for the plants. While using liquid fertilizer, the instructions are to be followed for mixing and application in the pot. Application of fertilizer is to be avoided when the plants suffer from wilting or heat stress. Recommended quantity of liquid fertilizer are to be applied by diluting with water. Then the solution is applied in the pot or the container with plant. Manure should be applied at the time of preparation of pot mixture. Subsequently, regular application of manure and/or fertilizer may be done depending on the type of potting media and rate of plant growth. Most of the liquid fertilizers provide nutrient to the plant for about one to two weeks. The soil in the potted plants needs to be changed in every 12 to 18 months.

As the potted plants have limited space for root growth, excess quantity of fertilizer application results in high concentration of chemical fertilizer in the potting medium. It may cause damage to the root and ultimate plant growth. Over-fertilization in potted plants may exhibit crust formation on the soil surface, yellowing of lower leaves, browning leaf margin, defoliation and death of young plants. There may be detrimental effect of fertilizer on the potted plants due to following reasons.

- Application of excess quantity of fertilizer at a time
- Frequent application of fertilizers with short time interval
- Improper and imbalance use of fertilizer
- Poor drainage system in the pot or container
- Exposure of the pot or container to high temperature
- Moisture stress conditions in pot or container
- Application of fertilizer on leaf or shoot may result in scorching effect

Bonsai

Bonsai is the practice of growing plants in miniature form by retaining its natural look. Now-a-days, there is increasing demand for growing bonsai plants of various species in the household. Almost any tree or shrub can be turned into a bonsai with proper care and management. The success of bonsai plant depends on shape and care of the plants chosen. The basic principle involves pruning of the roots and branches/foliage of the plant to obtain a dwarf form. Bonsai plants are grown in 5 to 7.5 cm deep shallow flat pots. Round, oval, hexagonal or square shaped pots with 25 to 30 cm diameter are preferred for raising bonsai plants. Colourful, designer containers made up of clay, stone, terracotta or ceramic give the bonsai plant an attractive look. The potting mixture is to be prepared by mixing sand, soil and compost/ FYM for proper growth of the plant. The media should have good water holding capacity for proper plant growth. Media can be covered with moss and sand pebbles to give a natural look.

Plants with small leaves, short internodes, attractive bark and good rooting habit are suitable for bonsai. These plants should sustain intense pruning of root or branch to get desired shape. Plants with large leaves do not turn out to be attractive as the leaf size doesn't reduce significantly in the bonsai process. Plants with long internodes may not have enough branches as the internodes do not shorten significantly during the dwarfing process. Banyan, *peepal,* lemon, bougainvillea, bottle brush, casuarinas, orange, mango, sapota, jackaranda, neem, pine, silver oak, etc. are suitable for bonsai preparation.

Basic Principles

- The bonsai plant should retain its natural look
- The plants should be suitable to local climatic conditions
- The bonsai plants should be placed in an airy place
- Adequate sunlight may be available for the growth of bonsai plant
- Regular formative pruning is to be done to create an attractive and uniform crown
- Provision of balanced nutrition through organic and/or chemical sources
- Regular watering of the plant should be ensured
- The bonsai plant may be protected from harmful insects, diseases and other problems

Steps of Bonsai Making

There are several steps required for preparation of bonsai plant. Some of those are described here.

***Pruning*:** Pruning is the primary operation in bonsai preparation to keep the plants in miniature form. Good quality and sharp pruning scissors/cutters are used to cut the root and shoot of the plant. The best time to prune the plant is during rainy season. While pruning the plant, branches with uneven growth are to be removed. Root pruning is important to get a good bonsai plant. However, root pruning is not done in the first season. The root system should be pruned after proper establishment of the plant.

***Wiring*:** Wiring is required to give the plant desired shape. Aluminium or copper wire is used for this purpose. The plant is wrapped carefully around the branches to give the plant desired shape. Wiring can be done round the year. The wire should be removed before it starts getting embedded in the plant body.

***Manuring*:** The bonsai plants need to be supplied with plant nutrients to meet the nutritional requirement of the plants grown in small pots with limited space. This can be done by application of organic manure, water soluble fertilizer or liquid fertilizer to keep the plant healthy.

***Watering*:** Bonsai plant should be watered regularly. But, over-watering may result in root rot and wilting of the plant. Water is applied to the pots based on the type of plant and growth media used.

***Placement*:** The potted bonsai plant is to be placed at indoor or outdoor depending on its natural growing habit. Plants accustomed with the local environment are to be placed in outdoor. In case of bonsai placed indoors, intermittent exposure to outdoor environment is required to keep the plant healthy.

***Potting and repotting*:** Most of the bonsai plants need to be repotted every 2-3 years to put those in fresh growing media and to trim the root. During repotting, one-third of the roots and excessive branches are to be pruned. To enhance the age of the plant, one-third of the root is kept exposed above the pot mixture. Care should be taken to avoid shading of the lower branches by the upper ones. The twisted and tangled roots are to be straightened before repotting. Repotting is to be done preferably in rainy season.

Types of Bonsai

There are several types of bonsai depending on the plant and local situation.

Formal upright: It is the most natural style with straight trunk. One-third of the branches from the bottom are removed and other branches are drawn downward.

***Informal upright*:** In this case, the trunk is lightly curved. This can be achieved by using wire or cords.

***Broom*:** It has a thick and finely branched crown. The trunk is straight and divided into several branches.

***Slanting*:** In this case, the slope of the trunk is very prominent. The branches are horizontal or slightly drooped down. The surface roots are unstable in appearance with well anchored impression.

***Windblown*:** In this case; the trunk, branches and twigs are trained in a single direction to give the effect of a strong wind and storm.

***The clasped-to-stone*:** It is a difficult style for bonsai making. The plant is shaped according to the shape of the rock or gravel.

Things to be Avoided

- Branches should not cross each other
- Branches should not emerge from opposite sides of the trunk at same height
- There should not be any reverse taper in the canopy
- Avoid plants with large leaves, flowers and fruits
- Do not allow the plant to enter into reproductive phase before the bonsai takes shape
- Do not wire the plant too tightly, which may damage the bark
- Do not use blunt cutting tools
- Do not go for extreme dwarfing of the bonsai, where its natural characteristics are not visible

Lawn

Lawn is a grass covered area to form a green carpet in the landscape. Lawns have become an integral part of landscaping in urban areas. A patch of green grass enhances the beauty of the garden including adjacent building. A lawn adds to the beauty of the background for trees, shrubs and flower beds. Location of the lawn is decided as per layout of the garden. Lawn should be well exposed to direct sunlight. The site preparation for developing a lawn includes tilling the soil, levelling the land and addition of organic manures. In heavy soil, coarse sand can be added to a depth of 20 cm for obtaining desired soil texture. The bed should be free from weeds, especially nut grass. The slope of the lawn should be well maintained to facilitate drainage.

Cynodon dactylon or *doob* grass is the most common grass used in lawn preparation. It is a dark green grass with fast spreading habit. Kentucky blue grass (*Poa pratensis*) , wood meadow grass (*Poa nemoralis*), *Agrostis tenuis, Festuca rubra*, Bermuda grass, etc. are also used for lawn making

The cheapest method of lawn establishment is through dibbling of small pieces of grass roots. The root cuttings are dibbled at a spacing of 10-15 cm in a levelled ground. The plot should be provided with light irrigation before planting of roots. It takes nearly six months for the grass to completely cover the area. A good quality lawn can be developed by regular weeding, mowing, rolling and watering. Application of excess water should be avoided to maintain the greenery of the grass.

Turf plastering is another method for development of lawn. In this method, chopped grass pieces of 5-7 cm length are mixed with garden soil, fresh cowdung and wood ash. Required quantity of water is added to form a thick paste. This mixture is spread evenly at a thickness of 2.5 cm on the surface of the area marked for lawn preparation. Prior to planting of grass cuttings, the area should be well prepared with addition of organic matter and light irrigation. The ground should be rolled next day to allow the grass to spread evenly.

Regular rolling, mowing, watering and restoration of patchy places should be done to maintain a good quality lawn. Application of well decomposed compost @ 10 kg per 10 m^2 area along with 200 g urea in 10 m^2 area thrice during a year helps to maintain the greenness of the grass. Regular weeding is essential to keep the lawn clean and presentable. Some of the common weeds found in lawns are nut grass (*Cyperus rotundus*), kansa (*Saccharum spontaneum*), thatch grass (*Imperata cylindrical*), water weeds (*Oxalis corniculata*), etc. It is difficult to eradicate *Cyperus rotundus* due to its deep root system and perennial growth habit. Manual weeding by pointed *khurpi* is generally practiced in lawn.

Mowing is done at regular intervals to discourage reproductive stage of weeds and restrict seed production. Mowing also helps in maintaining uniformity of the grass carpet. Cutting the grass too short can inhibit growth of root system, which may provide space for weed growth. The ideal height of grass in lawn is 8 to 10 cm. Mowing is done when the grass is dry. Raking the lawn once a year, prior to monsoon with addition of organic manure accelerates vigorous growth of the grass. The best time to water the lawn is morning hours as there is minimum evaporation loss of water. Care should be taken to apply water uniformly across the lawn avoiding excess flooding or missing certain spots.

Lawn grass is prone to pest and disease infestation. Some of the common insects affecting lawn are grub, cut worm, leaf hopper, etc. The grubs cut the root of grass, which results in brownish dead patches. It can be controlled by application of neem oil cake at the time of land preparation or chloropyriphos @ 2ml/lit of water at appearance of the symptom.

Cut worms damage the leaves resulting in dead spots. Spraying of profenophos @ 2 ml/litre of water can check cut worm attack. Diseases like brown patch, leaf spot and powdery mildew are commonly seen in lawn grasses. Brown patch and leaf spot diseases can be controlled by propiconazole or mancozeb. Sulphur dusting can control powdery mildew disease. Alternatively, neem oil or neem based pesticides can be used to control insect and disease infestation.

Baby Corn

Maize is cultivated not only for grain use, but also for various other purposes like quality protein maize, baby corn, sweet corn, popcorn, waxy corn, high oil corn, etc. Maize varieties grown for special purpose are collectively known as 'Specialty Corn'. Among these, baby corn and sweet corn are gaining popularity.

There is increasing demand for baby corn in the urban areas. Baby corn is gaining popularity for household consumption and commercial use. Baby corn is the young cob harvested just after emergence of white silk. It is used for preparation of salad, soup, pakoda, curry and other dishes. As there is growing demand of baby corn in the urban areas, it can be grown in the backyard garden to meet the requirement of home consumption. It is easy to grow baby corn in home gardens. It requires comparatively cooler climate for quality production. It can be suitably grown during winter season. However, in places of high altitude, it can be grown throughout the year. Adequate sunlight with temperature range between 22 and 28 degree Celsius is suitable for growing baby corn. It cannot perform well in areas receiving higher temperature. This crop can be grown in well drained loamy soil with high fertility. The optimum pH range suitable for good crop growth is 5.8 to 7.0.

There are several varieties available for baby corn cultivation. Some of those are VL Baby Corn-1, Kiran, Vivek Hybrid 27, HM-4, etc. Due to closer spacing of plants, the seed requirement is about 25 kg/ha. About 250 gram seed is required for sowing in 100 m^2 area in home gardens. The seeds need to be treated with thiram @ 3 gm or carbandazim @ 2gm per kg of seed to protect the crop from various fungal diseases. Seeds can also be treated with microorganisms such as *Pseudomonas fluorescens* @ 10 gram or *Trichoderma viridae* @ 4 gram per kg of seed.

Land should be prepared thoroughly to make the soil pulverised. Well decomposed farm yard manure or compost may be applied in the soil during land preparation. In case of backyard farming, 50 kg farm yard manure or compost can be applied in 100 m^2 area. The seeds are sown at a spacing of 40 cm x 20 cm to accommodate nearly 1250 plants in 100 m^2 area. Before sowing, lines are drawn with the help of trench hoe at a distance of 40 cm from line to line. For 100 m^2 area in home gardens, 1200 gram nitrogen, 600 gram phosphorus and 600 gram potash are to be applied. The basal fertilizer consisting of all phosphorus & potash and half of nitrogen may be applied at the time of sowing. The seeds are put in the lines at seed to seed distance of 20 cm and covered with soil. Light irrigation is to be provided to facilitate germination of the seeds. Hoeing should be done in the field after three weeks of sowing, which is followed by earthing up operation. It helps in weed management and facilitates irrigating the crop. Half of nitrogen may be top dressed just before earthing up. Depending on soil moisture status, irrigation can be done at regular intervals to maintain crop growth. The tassels are to be removed after emergence for diversion of nutrient flow towards cob formation. Detasseling can be done by removing the tassel of the plant as soon as it emerges from the flag leaf. Leaves should not be damaged during detasseling operation.

Stem borer is a serious insect affecting baby corn crop. This insect affects the plant at early stage of growth. Sometimes, the crop is affected by leaf spot and sheath blight diseases. Line sowing, application of neem or karanja oilcake and keeping the field weed free can reduce the risk of disease and pest infestation. The insect and disease can be controlled by application of neem oil or neem based pesticides. Generally, chemical pesticides are not used in baby corn crop.

Time of harvest plays an important role in obtaining better quality baby corn. The crop is harvested just after emergence of white silk. Depending upon local climatic situation, it takes nearly 45-50 days during kharif and

55-60 during rabi season to get the first harvest. The quality of baby corn is adversely affected with delay in harvesting. Cobs are to be removed from the plant without disturbing the leaves. Then those are taken to a shade area for dehusking the cobs. Delay in dehusking may adversely affect the quality of baby corn. Normally, harvesting is done 2-3 times in a week. After harvesting of baby corn, green plants are used as fodder for cattle. About 10-12 quintals of baby corn and 220 quintals of green fodder is obtained per hectare with proper management practices. For home gardens, about 10-12 kg of baby corn is harvested from an area of 100 m^2.

Sweet Corn

Sweet corn is the product obtained from specialized varieties of maize for consumption as green cob. Sweet corn has consumer preference in urban areas due to high sugar content. It is used as human food in the soft dough stage, when the grain is succulent and soft. Higher content of water-soluble polysaccharide in the grain adds texture and quality. Sweet corn is mainly produced for human consumption either as fresh or processed product. Sweet corn can be grown in urban areas with little effort. Seeds of sweet corn contain a sweetish starch and are characterized by wrinkle appearance on drying. Sweet corn can be cultivated during all the seasons. However, assured irrigation facility is required for growing during winter and summer seasons. This crop can be grown in different types of soils. However, loamy soil with high fertility level is suitable for this crop. Well-drain condition is required during *kharif* and the soil should have better water holding capacity for cultivation during *rabi* and summer seasons. The optimum pH range suitable for sweet corn is 5.8 to 7.0. Several varieties are available for cultivation of sweet corn. Prominent among those are Madhuri, Priya, Win orange, Shalimar Sweet Corn-1, VL Sweet Corn Hybrid-2, Hi-Brix 53, etc. As the seeds are shrunken with light weight, it requires comparatively less quantity of seed for sowing. About 10-12 kg of seed is sufficient to sow in one hectare area. For a small patch of 100 square meter, it is necessary to sow 100-120 gram of seeds. Seeds can be treated with three gram thiram or two gram carbendazim per kg of seed to protect the crop from seed borne diseases. Alternately, seeds can also be treated with 10 gram of *Pseudomonas fluorescens* or 4 gram of *Trichoderma viridae* per kg of seeds.

Sweet corn requires good soil preparation for better plant growth. The soil should be made well pulverised before sowing. Land can be prepared by digging with spade in the kitchen garden. Farm yard manure or compost may be added in the soil at the time of final land preparation. About 50 kg of farm yard manure or compost is required for a small plot of 100 m^2 area. Uniformity of plant distribution plays a vital role in production of sweet corn. Evenly spaced plants have greater yield potential than unevenly spaced stands. Sowing is to be done in lines at a row spacing of 60 cm. Lines are to be drawn with the help of plough or trench hoe. Seeds are sown at a spacing of 25 cm in the lines.

Sweet corn yield is largely affected by nutrient application. Fertilizer is to be added as per result of soil test. In home gardens; basal application of 300 gram nitrogen, 600 gram phosphors and 60 gram potassium is to be done for an area of 100 m^2. The crop can respond to higher level of fertilizer application in case of hybrid.

Weeds are to be controlled for suitable growth of the crop. Hoeing should be done at three weeks after sowing for control of wseeds. It should be followed by earthing up operation for facilitating drainage and irrigation. Top dressing of nitrogen @ 600 gram per 100 m^2 area can be done just before earthing up operation. Second top dressing of nitrogen is done at initiation of tassel with a dose of 300 gram per 100 m^2 area. Depending on soil moisture status, irrigation is to be provided at regular intervals to maintain good crop growth. During *rabi* or summer season, light irrigation is to be given after sowing for early germination. The crop requires 6-7 irrigations depending on the type of soil. Tasseling, silking and grain filling stages are critical periods for irrigation requirement.

The crop suffers due to infestation of insects like stem borer, cob borer and aphid. Stem borer affects the plant at early stage with symptoms of whorls on the leaf, wilting of central leaf and puncture mark on the stem. Cob borer affects the young cob making those unfit for consumption. Aphids suck the cell sap resulting in stunted growth of the plant. These can be controlled by application of neem based pesticide and neem oil. Besides, profenophos can be used to control stem borer and cob borer. Aphids can be controlled by application of imidachloprid. This crop is also affected by diseases like leaf spot, sheath blight and turcicum leaf blight. These diseases can be controlled by application of propiconazole or carbendazim or mancozeb. Clean cultivation and line sowing at proper spacing can reduce the incidence of disease and insect in sweet corn. Addition of neem or karanja oilcake in the soil at the time of land preparation prevents disease and pest infestation.

Harvesting should be done at right time to obtain good quality sweet corn. Normally, the optimum time of harvest is about 25-30 days after silking. This may vary depending on season, variety and management practices. The green cobs are harvested when the silk is completely dry, but husk colour is green and the kernel is soft. This is adjudged by pressing the kernel with nail. If harvesting is delayed, the quality of sweet corn is reduced and it loses sweetness. About 670 cobs can be obtained from an area of 100 m^2. Immediately after harvest, the cobs are taken to shade area to maintain the quality. The cobs are to be consumed as quick as possible. Storage for long period may result in loss of sweetness of the cobs.

Organic Manure Production

Now-a days, there is demand for organically grown agricultural produce for environment protection and health related issues. It has become a necessity to get organic food in urban households. In backyard gardening, people prefer to use organic inputs for growing crops. There is strong need to prepare organic material in the home garden to recycle the biodegradable waste and minimize the cost incurred for procuring fertilizer and manure. Organic manure can be prepared in many ways for converting the discarded household waste to useful manures. This can be achieved through preparing compost or vermi-compost in the backyard by utilizing the bio-degradable waste materials produced from the garden and household.

***Compost preparation*:** Composting is a simple method of recycling the household waste through decomposition of garden and kitchen wastes. Composting can be done at different places such as backyard, terrace or rooftop. Composting should preferably be done in outdoors to get better result. Various waste materials such as vegetable peels, fruit peels, wasted cooked food, dried leaves, garden wastes, etc. are used for preparation of compost. These materials are put in to a pit in the backyard or composting bins available in market. Any container like a bucket or a normal dustbin or a garden pot can also be used for this purpose. About 4-5 holes are made at different levels on the container to facilitate aeration. A plastic tray is kept under the container to hold any spills. The bottom most layer of the container is to be filled with soil. After the container is filled with waste materials, it is covered with another layer of soil at the top. To fasten the process, semi composted material can be added to compost bin. The pile of waste is to be turned with a garden rake at 4-5 days interval to facilitate aeration and quick decomposition of the waste materials.

***Vermi-compost*:** Vermicomposting is the process of converting the organic waste into organic manure by using surface dwelling earthworms like *Eudrilus eugeniae, Lumbricus rubellis, Eisenia foetida, Peronyx excavates,* etc. Vermi-compost is a rich source of nutrients required for plant growth. All types of biodegradable wastes such as garden waste, household waste, kitchen waste, leftover food, etc. can be used for vermi-compost preparation.

Vermi-compost is to be prepared in any place having shade. A thatched or polythene covered roof may be provided to protect the vermi unit from direct sunlight and rain. Vermi-compost is prepared in a cemented tank of 75 cm height, 90 cm width and of convenient length. The slope is maintained at the bottom of the tank to drain the excess water. Alternatively, vermi-compost can be prepared in wooden boxes, plastic buckets or in any container to meet the requirement of small household gardens. A layer of soil or sand with 5-7 centimetre depth is placed at the bottom of the tank. Partially decomposed cow dung and biodegradable wastes like kitchen waste and field waste are placed layer-wise up to a depth of 20-30 centimetre. The earthworms are released over the mixture and covered with dry straw or gunny bag. Water is sprinkled on it regularly to maintain the moisture content of the compost. Precautions are to be taken to protect the compost tank from attack of ants, lizards, mouse, snakes, etc.

After decomposition, the castings of the earth worm formed on the top layer of the tank are collected at weekly intervals. The castings are scooped out and heaped under the shade. Periodical harvesting is necessary for free flow of earth worm and improving quality of compost. After harvest, vermi-compost can be stored in a dark and cool place. Sprinkling of water may be done to maintain adequate moisture.

Vermi-wash :Vermi-wash is a brown coloured liquid manure suitable for foliar spray in plants. It is prepared by collecting the water, which has passed through a worm culture column. It is a rich source of plant nutrients and various beneficial microorganisms. Vermi-wash can be produced in the units established by using empty barrels, buckets or small earthen pots. A small pot or bucket is taken and a tap is attached at the bottom. The lower most part is filled with pebbles. The next layer is filled with coarse sand for easy percolation of water. Then a layer of cow dung and soil mixture (1:1 ratio) is added on it to fill up the pot. The earthworms are left on this layer. The earthworms will go inside the pot and get their food from the mixture of cow dung and soil. After installation of the unit, the tap is kept open for 48 hours for flow out of water along with the impurities. Vermi-wash is collected through the tap after 10 days. It is sprayed on the garden plants or potted plants by diluting with water to provide nutrition.

***Pot manure (haandi khata)*:** This is an organic formulation used to provide plant nutrients and control various types of disease & pest of plant. This also helps to improve the hormonal balance of the plants. Cow dung, cow urine and gur (jaggery) are placed in the earthen pot and mixed thoroughly. Neem, *Arakha* and *Karanja* leaves are chopped into small pieces and added to the pot. The pot is kept under shade for a week and stirred intrmittently. Then the liquid is decanted from the pot. This is sprayed on the plants by diluting @ 15-20 ml per litre water.

Green manure**:** Green manuring can be done to improve the soil structure and fertility of the garden. This can be done by incorporating the plant residues and green plants into the soil by ploughing or digging with spade. Plants like *dhaincha* (*Sesbania aculeata*), sunhemp (*Crotolaria juncea*), cowpea, guar, etc. can be used for green manuring. The green manuring crops should be of short duration and produce foliage profusely. Those should be succulent enough for easy decomposition. Adoption of green manuring may minimize the nutrient requirement of garden plants and helps in better root growth.

Mushroom Production

Now-a-days mushroom has become a preferred food item in daily diet. It has high nutritional value and several health benefits. Mushroom can be grown in diverse climatic conditions with proper management practices. It is an unique agri-enterprise, which provides return within a short span of time with minimum investment. The return per unit area is comparatively higher than other agri-based enterprises.

Mushrooms are a group of spore bearing fruiting fungi, which derive food from various media such as wood, soil, straw and other materials. There are about 50,000 reported species of fungi, out of which about 10,000 comes under edible category. Mushrooms are often used as food although some species are poisonous in nature. Mushroom is rich in protein, carbohydrate and vitamins. Mushroom contains all the essential amino acids and vitamins like thiamine, riboflavin, niacin, pantothenic acid, biotin, folic acid, etc. Since mushrooms possess low caloric value with high in protein, fibre and K: Na ratio; those are ideally suited for diabetic and hypertension patients.

The varied agro-climatic conditions along with abundant availability of agricultural waste and manpower provide ample scope for cultivation of various types of edible mushroom in India. It can be a boon for the landless person, unemployed youths, home makers and other entrepreneurs to earn substantial income. It requires less land as compared to other agricultural enterprises and can be grown in backyard, rooftop, terrace or balcony. It utilizes the household and agricultural wastes and thereby helps in minimizing environment pollution.

Mushroom cultivation has important role to play in ensuring nutritional security of the household. It can augment income generation and self-employment opportunities for home makers, youth and senior citizens in urban areas. Mushroom cultivation can be a good recreational activity for the members of urban families, which help in effective utilization of leisure hours and supplies additional nutrition in terms of food. In the present day context, many home makers are interested to produce mushroom for their family consumption by utilizing the space available in and around their house. Mushroom growing needs minimum investment and simple management technique for which there is low risk for trying this even by agriculturally novice person. Mushroom can be cultivated in a place having good aeration and protected from direct sunlight or rainfall.

The cultivation techniques are standardized for about twenty mushrooms and some of these have been recommended for commercial cultivation. However, only four types of mushroom are preferred for cultivation. Those are paddy straw mushroom (*Volvariella spp.),* oyster mushroom (*Pleurotus spp.),* button mushroom (*Agaricus spp.)* and milky mushroom (*Calocybe indica).*

Paddy Straw Mushroom

Paddy straw mushroom (*Volvariella volvacea*), also known as straw mushroom, is widely cultivated in various states. This mushroom is grown in the tropical or sub-tropical climate. It is fast growing, easy to cultivate and has wide acceptability by the consumers. The raw materials required for paddy straw mushroom cultivation are cheap and abundantly available across the country. High temperature requirement for its cultivation also makes it a good choice during summer and kharif seasons. A single bed of paddy straw mushroom requires one square meter area. Hence, four beds can be taken up in a terrace of six square meter area, which can produce around 6 kilograms of mushroom per month. This is adequate to meet the requirement of the family of five members. However, the beds are to be prepared at certain time interval to obtain continuous production.

Nutritive value: Paddy straw mushroom is considered as a healthy food. The unique flavour and attractive textural properties make this mushroom stand out among other edible mushrooms. The nutritive value of paddy straw mushroom is influenced by the growing method and harvest stage.

Substrate required: Paddy straw mushroom requires high cellulose and low lignin containing substrate for good production. Cultivation of paddy straw mushroom can be done by using various substrates like paddy straw, water hyacinth, banana leaves, cotton waste, sugarcane bagasse, maize stover, etc. Commonly cultivated varieties of paddy straw mushroom are *Volvariella volvacea, Volvariella diplasia* and *Volvariella esculenta.*

Cultivation practices: This can be adopted in the rooftop or open space adjacent to the building. In this method, beds of 100 cm X 100 cm size are prepared with the help of bamboo or wood, which rest on brick/stone platform.

The bed size may vary depending on the season, type of spawn used and local climatic condition. The beds are kept under the shade to protect those from rain or direct sunlight. One bed of paddy straw mushroom needs about 7 kg freshly harvested paddy straw, one spawn bottle and 250 gram of gram/dal flour.

Depending on availability of straw, the ideal length of straw should be about 70-80 cm. The leaves are removed from the straw and the straw is made to bundles of 7-8 cm diameter. These bundles are soaked in clean water for 6-8 hours in a cemented or polythene lined water tank. Later, bundles are taken out and placed on raised platform to drain out excess water.

- A frame, prepared out of bamboo or wood is laid on brick or stone foundation at a height of 12-15 cm from the ground level.
- The required paddy straw, spawn and gram/dal flour are divided into four equal parts.
- One part of paddy straw is placed on the platform with tip of the straw facing one direction.
- One fourth of the spawn is placed on the side of the beds leaving a margin of 5-8 cm from edges. One fourth of gram/dal flour is placed on the mushroom spawn.
- Another set of bundles of straw are to be placed on the bed in the direction opposite to the prior layer.
- Another part of the spawn is placed on the straw at about 5-8 cm from the edges of the layer bundles.
- One part of the powdered dal or gram is placed on the straw.
- Likewise, third layer of the straw is placed on the bed with a direction opposite to the previous straw layer.
- Then, two parts of spawn and two parts of gram/dal flour are distributed on the paddy straw.
- Then, the rest part of paddy straw is placed on the bed. The straw is to be put in the direction opposite to the previous layer of the straw.
- Finally, the bed is covered with a transparent polythene sheet to maintain desired temperature and humidity.
- Mushroom mycelium is completely spread in the bed within 7-8 days of spawning and the plastic cover is removed at this stage.

- Water is sprayed on the bed with the help of sprayer at least once in a day depending on the moisture content of the bed.
- Mushroom fruiting occurs after spawning at favourable moisture and temperature condition.
- Fruiting continues for about 10-12 days and yield of 1.0-1.5 kg mushroom can be obtained per bed. In some areas, second phase of fruiting is also observed.
- Harvesting of mushroom is done prior to complete opening of the cap.
- Paddy straw mushrooms are very delicate in nature and can be stored under refrigerated condition for 2-3 days.
- After harvest of mushroom, the substrate can be used as manure in the field after composting. It can be an excellent base material for vermicomposting.

Oyster Mushroom

Oyster mushroom (*Pleurotus spp.*) is commonly known as *Dhingri* mushroom in India. The oyster like shape gives it the name of oyster mushroom. Its cultivation can be done on various substrates derived from agricultural and organic waste materials. Some of the commonly used substrates include straw of different cereals, sugarcane bagasse, cotton waste, saw dust, maize stover, etc. Commonly cultivated species of oyster mushroom are *Pleurotus sajor-caju, Pleurotus ostreatus, Pleurotus florida, Pleurotus eryngii, Pleurotus flabellatus, Pleurotus opuntiae, Pleurotus platypus, Pleurotus pulmonarius,* etc. These species are grown under different agro-climatic conditions. For this type of mushroom, 22-26°C temperature and relative humidity of 80-85% is suitable. In urban households, oyster mushroom can be grown in the balcony, rooftop, terrace or in the backyard. About10 bags of oyster mushroom can be produced in terrace of six square meter area, which can produce around 10 kilogram of mushroom per month.

***Cultivation practices*:** It is very simple to grow oyster mushroom for home consumption. Oyster mushroom is produced in bags by using straw. The bags can be either hanged in the balcony or placed on a frame in the backyard/roof top. A bag of oyster mushroom requires about 2 kg of freshly harvested paddy or wheat straw, one spawn bottle and a polythene bag of 50cm X 75cm size.

- Oyster mushroom is commonly cultivated in areas having ample availability of rice straw or wheat straw.
- The outer leaves are removed from the straw and the straw is cut into small pieces of 4-6cm length.

- Soaking of straw is done for 6-8 hours followed by sterilization of straw by dipping in hot water at 80°C for 2 hours. Hot water treatment makes the substrate soft so that growth of the mycelium takes place easily. This method is suitable for small scale cultivation.
- Then the straw is taken out and excess water is drained off.
- The straw and spawn are divided into four equal parts.
- One end of the polythene bag is tied. Holes of about one cm diameter are made on the polythene bag at 10-15 cm distance for diffusion of gases and heat generated inside.
- One part of paddy/wheat straw is placed in the polythene bag and pressed down.
- Then one part of spawn is placed on the straw leaving a margin of 4-5 cm from the edge.
- Another set of straw (one fourth) pieces are to be placed in the bag.
- Another part of the spawn is placed on the straw at about 4-5 cm away from the edges.
- The third layer of the straw is placed in the polythene bag on which rest of the spawn is spread.
- Then the balance quantity of straw is placed on the top of the bag and pressed thoroughly.
- In another method, all the straw pieces and spawn are mixed and put inside the polythene bag.
- The bag is tied and hanged or placed on the frames, preferably in a dark place.
- Within 15-18 days of spawning, the white mycelium spreads in these bags. The straw pieces are made to a bundle like structure.
- Then the straw bundles are taken out of the polythene bag and kept in lighted place either by hanging from roof or placed on racks at a distance of 25-30 cm.
- Water is sprayed over the bags in the morning and evening hours to maintain desired moisture level.
- Pinning starts in next 4-5 days and fruit bodies become fully grown within a week of pinning.

- The mature mushrooms are harvested by slightly twisting and lifting.
- The average yield of oyster mushroom is around 1.5 to 2.0 kg per bag.
- The mushrooms are packed in perforated polythene bags in different packing after proper cleaning.
- These are either consumed or sold fresh in the market or stored in a refrigerator for 4-6 days.

Milky Mushroom

Milky mushroom (*Calocybe indica)* can be grown with fresh straw. Fresh straw (paddy/wheat) is chopped and soaked in clean water for 6-8 hours and subsequently soaked in hot water (80-90°C) for 40 minutes to achieve pasteurization. Space requirement of milky mushroom is similar to that of oyster mushroom.

Spawning is done when the substrate moisture level is at 62-65 %. After spawning, bags are kept in the dark place at a temperature of 35°C and relative humidity above 80 %. After 15-20 days, the substrate is fully covered with mycelium and bags are ready for casing. Casing is the process of covering the top surface of mushroom bags with pasteurized casing mixture at a thickness of about 2-3cm. Casing mixture is prepared by mixing soil and sand in the ratio of 3:1 and treated with formaldehyde solution (4%) about one week before use. At the time of use, casing mixture should be free from formalin smell. Fruiting starts after 10 days of casing at a temperature of 30-35°C along with relative humidity of 80-90 % and adequate light during the day time. The pin shaped mushroom matures in about a week. Mushrooms with 7-8 cm diameter caps are harvested by twisting. Those are cleaned and packed in polythene bags for marketing.

Button Mushroom

Now-a-days, there is increasing demand of consumers for button mushroom. Button mushroom is also known as white button mushroom (*Agaricus brunnescens*) that includes white, brown and cream variety. This mushroom is thickly fleshed and robust with thin gills on the underside of the cap. Button mushroom contains about 3.92 % protein, 1.09 % crude fibre, 1.25 % ash and 0.19 % fat. The environmental factors like temperature, relative humidity, light and air flow in the cropping room play a vital role in production of button mushroom. Button mushrooms are grown under controlled conditions. It requires 20-28 degree Celsius for vegetative growth and 12-18 degree Celsiusfor formation of fruit body. Well ventilated place with relative humidity of 80-90% is suitable for button mushroom production.

***Post-harvest management*:** After harvest, mushrooms have to be cleaned and graded either for immediate marketing or for further storage and value addition. The grading and sorting is done according to the colour, size, opening stage, length of the stem, damaged mushroom, etc. The mushrooms for fresh market are packed in plastic containers, perforated polythene bags or loose bags. Mushrooms can be stored in deep freezers or in vacuum freezers. Mushroom can be preserved in various ways for future use. Several value added products like *murabba*, pickle, soup powder, ketch-up, biscuits, candy, etc. can be prepared from mushroom.

Problems associated with mushroom production: During mushroom cultivation, some common problems may arise. Some of those are as follows.

- Poor growth of the fungi due to insufficient nutrient availability, improperly prepared substrate, too compact bed or poor quality spawn.
- Presence of contaminants because of no or poor sterilization of substrates or use of contaminated spawn.
- Mycelium drying out due to inadequate supply of water or excessive ventilation.
- Deficiency of light, use of degenerated spawn or too old spawn, excessively high temperature or poor ventilation result in poor fruiting.
- Death of young mushroom as a result of using degenerated spawn, infestation by insects, insufficient oxygen, abrupt temperature fluctuations or spread of fungal diseases.

Bee Keeping

Bee keeping or apiculture is the process of managing bee colonies with an objective to produce honey and other hive products. Apiculture can be practiced in backyard of urban households to get honey for home consumption. Maintaining a colony does not require big investment, land area or labour. Honey bees help in yield enhancement of garden plants by promoting pollination.

Honey bees live in colony. The honey bees are of three types such as queen, worker bees and drones. The workers are female bees, involved in taking care of the queen, collection of nectar & pollen, building comb, guarding the hive, keeping the hive clean and production of honey. The queen is the sexually developed female, which lays eggs to form drones and workers. A queen can lay up to 1000 to 1500 eggs in a day depending upon availability of food. The drones are male bees, responsible for fertilizing the virgin queen through mating. Drones are generally bulky as compared with worker bees.

In India, most commonly cultivated honey bee species are Indian Bee and Italian bee. Indian hive bee (*Apis cerana indica*) is commonly cultivated types with an average honey yield of 6-8 kg per colony per year. Italian bee (*Apis mellifera*) constructs parallel combs and produce 25-40 kg honey per colony per year.

Bee boxes are to be placed in areas having nectar and pollen producing plants in abundance. A clean source of water is required for meeting the requirement of bee colony and regulating the temperature of the hive. The area should be kept free from chemical pollution including agro-chemicals. The colonies should be provided with shade during summer season. The best time to start bee farming is the peak blooming period of plants in nearby areas. Bee colonies are to be procured from a reliable source.

Hives should be kept in light weight box made out of locally available seasoned wood (jamun, sal, etc.). The boxes are to be kept at a distance of three meters. Cleanliness may be maintained in the beehives by cleaning the bottom board, top cover, etc. The colonies are to be checked periodically for any abnormalities or changes in behaviour of the bees. It is not advisable to open the box and examine the colonies in cold, windy, cloudy and rainy days. While opening the box, smoker is to be used to decrease the aggressiveness of the bees. Protective dress and gloves are to be worn while opening the colony box. Crushing the bees during box inspection may irritate the bees leading to sting. The diseased and healthy colonies are to be handled separately.

***Equipment needed for bee keeping*:** Various equipment are needed for maintaining bee box. Wooden hive is used for bee keeping purpose. Frames of rectangular shape are put within a hive to enable the bees for constructing the comb. Gloves, bee suit, shoes and bee veil are used to protect the bee keeper from bee sting. Smokers are used to keep the bees calm during working in the hive. Some of the essential equipment required for bee keeping are hive tool, queen cage, queen gate, feeder and bee brush. Hive tool is used for movement of the frames in the bee hive. Queen cage is a handy and convenient tool used to transport the queen temporarily. A feeder (bottle, feeder plate & feeder frame) is used for feeding the honey bees. Bee brushes are used to separate the bees from the honey frame/comb.

***Care during dearth period*:** When the stored honey in the colonies is not adequate and nectar is not available in the area, sugar syrup is to be provided to the colonies. Feeding should be given after sunset. Old and dark combs are to be discarded. Before honey extraction, all the equipment, containers, etc. are to be washed thoroughly with warm water.

***Care during swarming*:** The bees start swarming due to non-availability of nectar/food or any type of unfavourable situation. When the swarming of the colony is noticed, the entrance gates of the colonies are to be closed with queen gate and necessary steps (like feeding) may be taken to discourage swarming.

Seasonal management of apiary: During summer season, the bee boxes are to be kept under shade. Wet gunny bags can be kept over the boxes to reduce the ambient temperature. Proper ventilation may be ensured by widening the entrance gate of the colony. Ensuring availability of fresh water near the box is helpful. During monsoon months, the boxes and the surrounding area are cleaned to prevent disease and pest infestation. Artificial feeding with sugar syrup and/or pollen substitute is to be done in case of unavailability of adequate quantity of nectar or pollens. Care should be taken to protect the bee boxes from attack of wasp, ant, frog, lizard, etc. With advent of winter season, the colonies are to be shifted to sunny places. The colonies are to be protected from cold winds.

***Management of pest*:** Honey bees may be affected by disease and pest. Wax moth, mites, wasps, birds, etc. cause harm to the bees and bee hives. The bee colonies are also affected by various fungal diseases. As chemical control of disease and pest is not adopted in bee keeping, management options are to be preferred to keep the colony healthy.

Selection of dry place with shade is suitable to install the boxes. Maintaining cleanliness near and in the beehives can prevent disease/pest attack. Selection of bee colonies from disease free brood and periodical inspection of the colonies for any abnormalities or changes in behaviour of the bees can minimize the risk of infection. Isolation of diseased colonies and handling those separately will be helpful. The contaminated equipment should be disinfected by dipping in soap solution. The equipment should be washed and dried before use. Harvesting is done carefully to get good quality honey. It is advisable to extract honey by honey extractor. Honey harvested by this method is free from impurities. Quality of honey varies depending upon the types of foraging plant from which the bees collect the nectar. Honey should be preserved in glass bottles after processing.

Selected References

Ahlawat, O.P. and Arora, B. 2016. Paddy Straw Mushroom (*Volvariella volvacea*) Cultivation. Published at: https://www.researchgate.net/ publication/343499927

Ahmed, S.J. and Sharma, S.C. 2001. Fruit and Vegetable Juice Therapy. Published by Pustak Mahal, Delhi.

Bonsai. https://indiaagronet.com/horticulture/CONTENTS/bonsai.htm.

Chadha, K.L. 2001. Handbook of Horticulture. Published by ICAR. New Delhi.

Chatterjee, A., Debnath, S. and Pal, H. 2020. Implication of Urban Agriculture and Vertical Farming for Future Sustainability. Published by Intechopen Limited.

Crop Production Techniques of Horticulture Crops. 2004. Directorate of Horticulture and Plantation Crops, Chennai and TNVU, Coimbatore.

Dimitri, C., Oberholtzer, L. and Pressman, A. 2016. Urban agriculture: connecting producers with consumers.British Food Journal.

Dulal, S. 2019. An Overview of Mushroom Farming, DO- 10.13140/RG.2.2.28189.87526

Honey bee farming information guide. 2018. https://www.agrifarming.in/honey-bee-farming-information-guide.

How To Keep Potted Plants Alive. https://patuxentnursery.com/blog/keep-potted-plants-alive/

How to prepare Organic Manure at Home. 2018. https://krishijagran.com/agripedia/how-to-prepare-organic-manure-at-home/

How to Start Bee keeping in Your Backyard. 2019. https://backyardbeekeeping.iamcountryside.com/beekeeping-101/how-to-start-a-honey-bee-farm/

https://agritech.tnau.ac.in/

https://agritech.tnau.ac.in/horticulture/horti_Landscaping_lawn%20making.html

Jain, S. 2017. Make Your Own Kitchen Garden: A Beginner's Guide. https://www.livspace.com/in/magazine/how-to-make-a-kitchen-garden-india.

Jones, J. Benton Jr. 2014. Complete Guide for Growing Plants Hydroponically. CRC Press Taylor & Francis Group, USA.

Kumar, H. Kumar, Anuj, Kumar, Suresh and Lalit Kumar. 2019. Lawn Establishment and Maintenance. https://krishijagran.com/featured/ lawn-establishment-and-maintenance/

Kumar, P. and Saini, S. 2020. Nutrients for Hydroponic Systems in Fruit Crops. In Urban Horticulture - Necessity of the Future. Edited by Shashank Shekhar Solankey, Shirin Akhtar, Alejandro Isabel Luna Maldonado, Humberto Rodriguez-Fuentes, Juan Antonio Vidales Contreras and Julia Mariana Márquez Reyes. http://dx.doi.org/10.5772/intechopen.82900.

Mougeot, L. J. A. 2000. Urban agriculture: definition, presence, potentials and risks. In Growing Cities, Growing Food: Urban Agriculture on the Policy Agenda. A Reader on Urban Agriculture (Eds N. Bakker, M. Dubbeling, S. Gündel, U. Sabel-Koschella& H. De Zeeuw), pp. 99–117. Feldafing, Germany: DSE/ETC.

Munshi, N.A., Dar, G. H., Ghani, M.Y., Kauser, Y. and Mughal, N. 2010. Button Mushroom Cultivation. Published by Division of Plant Pathology Sher-e-Kashmir University of Agricultural Sciences and Technology of Kashmir, Srinagar

Nugent, R. 2000. The impact of urban agriculture on the household and local economies. In Growing Cities, Growing Food: Urban Agriculture on the Policy Agenda. A Reader on Urban Agriculture (Eds N. Bakker, M. Dubbeling, S. Gündel, U. Sabel-Koschella& H. De Zeeuw), pp. 67–97. Feldafing, Germany: DSE/ETC.

Package of practices of flower crops. 2017. Published by Floricultural Research Station. Sri Konda Laxman Telanagna State Horticultural University, Hyderabad.

Ramachandrappa, B.K. and Nanjappa H.V. 2006. Speciality corns Pop corn, sweet corn, Baby corn. Kalyani Publishers, Ludhina. Pp 33.

Sinha, N.K., Sidhu, J.S., Barta, J., Wu, J.S.B. and Cano, M.P. 2012. Handbook of fruits and fruit processing. Published by John Wiley & Sons, Ltd.

Solankey, S.S., Akhtar, S., Maldonado, A.I.L., Rodriguez-Fuentes, H., Contreras, J.A.V. and Reye, J.M.M. 2020. Urban Horticulture Necessity of the Future. Published by Intechopen, London.

Subramanian, P., Thamban, C., Josephrajkumar, A., Hegde, V., Hebbar, K.B., Bhat, R., and Niral V. 2020. Coconut. Published by ICAR - Central Plantation Crops Research Institute Kasaragod.

Sumangla, H. P. Malhotra, S. K. and Chowdappa, P. 2016. Urban and Peri- Urban Horticulture -A Perspective. Published by Confederation of Horticulture Associations of India.

Sumangla, H.P., Malhotra S.K. and Chowdappa P. 2013. Urban and Peri-urban Horticulture-A perspective. Confederation of Horticulture Associations of India, New Delhi.

Thapa, C.D., Prakasam, V. and Singh, M. Mushroom Culture. www.AgriMoon.Com.

Venkatesh, S., Sujay, Rakshit and Sekhar, J.C. 2003. Sweet corn. *Speciality Corn Technical Series1*. Directorate of Maize Research, New Delhi.